REVOLUTIONS IN SCIENCE

Dawkins vs. Gould

Survival of the Fittest

Kim Sterelny

Series editor: Jon Turney

ICON BOOKS UK

TOTEM BOOKS USA

Published in the UK in 2001
by Icon Books Ltd., Grange Road,
Duxford, Cambridge CB2 4QF
E-mail: info@iconbooks.co.uk
www.iconbooks.co.uk

Published in the USA in 2001
by Totem Books
Inquiries to: Icon Books Ltd.,
Grange Road, Duxford,
Cambridge CB2 4QF, UK

Sold in the UK, Europe, South Africa
and Asia by Faber and Faber Ltd.,
3 Queen Square, London WC1N 3AU
or their agents

Distributed to the trade in the USA
by National Book Network Inc.,
4720 Boston Way, Lanham,
Maryland 20706

Distributed in the UK, Europe,
South Africa and Asia by
Macmillan Distribution Ltd.,
Houndmills, Basingstoke RG21 6XS

Distributed in Canada by
Penguin Books Canada,
10 Alcorn Avenue, Suite 300,
Toronto, Ontario M4V 3B2

Published in Australia in 2001
by Allen & Unwin Pty. Ltd.,
PO Box 8500, 83 Alexander Street,
Crows Nest, NSW 2065

ISBN 1 84046 249 3

Series editor: Jon Turney

Originating editor: Simon Flynn

Typesetting by Hands Fotoset

Printed and bound in the UK by
Cox & Wyman Ltd., Reading

Contents

Acknowledgements

Thanks to Mike Ridge, Matteo Mameli, Darryl Jones, Jochen Brocks, Dan Dennett and Jon Turney for their thoughts on a draft of this book. Thanks also to the Philosophy Program, Research School of the Social Sciences, Australian National University and the Department of Philosophy, Victoria University of Wellington, whose support made it possible for me to write it.

Dedication

For Peter: Friend and colleague

Part I
Battle Joined

1 A Clash of Perspectives

Science in general, and biology in particular, has seen its fair share of punch-ups. In the 1930s and 1940s, Britain's two greatest biologists, J.B.S. Haldane and R.A. Fisher, feuded so vigorously that their students (John Maynard Smith tells me) were hardly allowed to talk to one another. But their behaviour was civilised compared to the notorious feuds in biological systematics between cladists – notorious for wielding unintelligible terminology and vituperation in equal measure – and their opponents. Mostly these fights are kept more or less in-house, often because the issues are of interest only to the participants. Almost no one except systematicists are interested in the principles by which we tell that *Drosophila subobscura* is a valid species. But sometimes these disputes leak out into the open. Richard Dawkins and Stephen Jay Gould have different views on evolution, and they and their allies have engaged in an increasingly public, and increasingly polemical, exchange.

At first glance, the heat of this exchange is puzzling. For Dawkins and Gould agree on much that matters. They agree that all life, including human life, has evolved over the last 4 billion years from one or a few ancestors, and that those first living things probably resembled living bacteria in their most crucial respects. They agree that this process has been wholly natural; no divine hand, no spooky interloper, has nudged the process one way or another. They agree that chance has played a crucial role in determining the cast of life's drama. In particular, there is nothing inevitable about the appearance of humans, or of anything like humans: the great machine of evolution has no aim or purpose. But they also agree that evolution, and evolutionary change, is not just a lottery. For natural selection matters too. Within any population of life forms, there will be variation. And some of those variants will be a touch better suited to the prevailing conditions than others. So they will have a better chance of transmitting their distinctive character to descendants.

Natural selection was one of the great discoveries in Darwin's *Origin of Species* (1859). If a population of organisms vary one from another; if the members of that population differ in fitness, so one is more likely than another to contribute her descendants to the next generation; if those differences tend to be heritable, so the fitter organism's offspring share her special characteristics, then the population will evolve by natural selection. Australia is renowned for its poisonous snakes and of these the taipan is the most famously venomous. Let's consider the mechanism through which it became so impressively lethal. If a population of taipans differ in the toxicity of their venom; if the more venomous snakes survive and reproduce better than less venomous ones, then taipans will, over time, evolve more toxic venom. Gould and Dawkins agree that complex capacities like human vision, bat echolocation, or a snake's ability to poison its prey evolve by natural selection. And they agree that in human terms, natural selection works slowly, over many generations. Bacteria and other single-celled organisms whip through those generations at speed, and that is why drug resistance outpaces new drugs. But for larger, more slowly reproducing organisms, significant changes take tens of thousands of years to build.

Adaptive change depends on cumulative selection. Each generation is only slightly different from the one that precedes it. Perhaps, very occasionally, a major evolutionary change appears in a single generation, as the result of one big mutation. But the parts of an organism are delicately and precisely adjusted to one another, so almost all large, random changes are disasters. Adding a horn to a horse's head might seem to provide it with a useful defensive weapon, but without compensating changes to its skull and neck (to bear the extra weight) it would be not only useless but detrimental. So large single-step changes, Gould and Dawkins agree, must be very rare. The normal history of an adaptive invention is a long series of small changes, not a short series of large changes.

Yet Dawkins and Gould have clashed heatedly on the nature of evolution. In two notorious articles in *New York Review of Books*, Gould scathingly reviewed *Darwin's Dangerous Idea*, a work of Dawkins' intellectual ally Daniel Dennett. In 1997, there was a better tempered but no more complimentary exchange in *Evolution*, as they traded reviews of each other's most recent creation.

Dawkins and Gould are representatives of different intellectual and national traditions in evolutionary biology. Dawkins' doctoral supervisor was Niko Tinbergen, one of the co-founders of ethology. Ethology aims to understand the adaptive significance of particular behavioural patterns. So Dawkins' background sensitised him to the problem of adaptation; of how adaptive behaviours evolve in a lineage and develop in an individual. Gould, in contrast, is a palaeontologist. His mentor was the brilliant but notoriously irascible George Gaylord Simpson. The match, if it exists, between an animal's capacities and the demands of its environment is less obvious with fossils than with live animals. A fossil gives you less information on the animal and its environment. So it is tempting to suppose that the passion of these exchanges reflects nothing deeper than competition for the same patch of limelight, magnified by different historical and disciplinary perspectives. I think that suspicion would be misplaced, and it's my aim in this book to explain why. Despite real and important points of agreement, their clash is of two very different perspectives on evolutionary biology.

For Richard Dawkins, the fit between organisms and environment – adaptedness, or good design – is the central problem evolutionary biology must explain. He is most struck by the problem Darwin solved in his *Origin*: in a world without a divine engineer, how can complex adaptive structures come into existence? In his view, natural selection is the only possible answer to this question. Natural selection is the only natural mechanism that can produce complex, co-adapted structures, for such structures are vastly improbable. Hence natural selection plays a uniquely important role in evolutionary explanation.

Moreover, and most famously, Dawkins argues that the fundamental history of evolution is the history of gene lineages. The molecular biology of genes – the chemical details of their action, interaction and reproduction – is alarmingly complex. But fortunately Dawkins does not allow himself to be bogged down in these details, and we can follow his lead. He argues that the critical agents in life's drama must persist over long periods *precisely because* the invention of adaptation requires a long series of small changes. Hence the target of selection is a lineage that persists over many generations. Gene lineages and only gene lineages satisfy this condition. Genes are replicated: there are mechanisms that copied some of my genes into my daughter's genome; and those same mechanisms are capable of copying those same genes generation by generation. So genes form lineages of identical copies. These lineages can be very deep in time. You have genes that you share with yeasts and other single-celled organisms; organisms that have been evolving separately for billions of years. Perhaps with the exception of those organisms that clone themselves, organisms do not form lineages of identical copies. Reproduction is not copying. My daughter is not a copy of me. An organism disappears at the end of its life. But an organism's genes may not disappear. If that organism, or a relative carrying a similar complement of genes, reproduced, then copies of the organism's genes will persist. They may do so for many generations.

Moreover, the chance that a gene will be copied is not independent of the character of that gene. It is true that some genes are *silent*, and just seem to be hitching a ride. But often genes influence their own replication prospects. They do so most overtly by their influence on the characteristics (the *phenotype*) of the organism that bears them. So genes influence their own prospects of being copied. Dawkins conceives of the fundamental struggle of evolution as the struggle of genes in lineages to replicate. Moreover, success for one gene lineage can mean failure for another. Dawkins' opponents often portray him as a crazed reductionist, thinking that only genes matter in evolution. That is not his view. Organisms are important, but primarily as a weapon in the struggle between gene lineages. Gene lineages usually compete with other gene lineages by forming alliances. Rival alliances build organisms. Successful organisms replicate the genes in the alliance that builds them. Thus macaw-making genes which build macaws suitable to the bird's circumstances become more common over time. The conflict between two macaws for a safe hollow in which to nest influences evolution by determining which lineages of macaw-making genes will be represented in the next generation, and in what numbers. The ecological struggle between organisms to survive and reproduce is translated into differential success for the genes that build the organisms.

In short, for Dawkins, the history of life is a history of a mostly invisible war between gene lineages. The beautiful biological mechanisms that we see on so many natural history documentaries are the visible products of that war. They are its weapons. For rival gene alliances are engaged in a perpetual arms race. In human arms races, weapons improve over time. So too will biological weapons, though that improvement has from time to time been disrupted by catastrophic and unforeseeable changes to the battlefield: episodes of mass extinction when many species disappear. These changes may be caused by the geology of the earth itself, as continents divide, mountains erupt, seas and ice fields advance or retreat. And they may be caused by forces external to the earth: by impact or by changes in the sun's behaviour. But between these episodes, selection is omnipresent and effective, sifting gene teams, building adaptive improvements in the organisms that are their *vehicles*, as Dawkins puts it.

Gould sees the living world very differently. Life today is fabulously diverse. But many forms of life that used to dominate their environments are no longer with us. Gould is a palaeontologist, and so much of his professional life concerns extinction: from the spectacular extinction of the dinosaurs, pterosaurs and huge marine reptiles, to the less obtrusive, and yet in Gould's eyes more fundamental, extinctions of weird marine invertebrates 500 or more million years ago. The first multi-celled animals in the fossil record lived, then disappeared, 600 million years or so ago. This 'Ediacara fauna' is so enigmatic that there is debate as to whether they were animals at all. The fossils consist of the remains of frond- and disc-shaped organisms, and interpretations of these fossils vary widely; some think they are more like lichen than animals. After the Ediacaran disappearance, in the so-called Cambrian era 530 million years or so ago, the main modern lineages came into existence. *Arthropods* (insects, crabs, and their kin) evolved. So did *bivalves* (oysters, clams and the like) and *molluscs* (snails and their relatives). *Jellyfish* and *sponges* were around too, though they may have appeared a little earlier than the others. A horde of the different kinds of worms appeared. So too did the first *chordates*; our group. But at the same time, many other lineages came into existence, only to go extinct again. Extinction, and its causes, is one of Gould's fundamental concerns.

Dawkins is impressed by the power of selection to build adaptations. Gould is equally struck by conservative aspects of the history of life. In their most fundamental ways, animal lineages do not seem to change over enormous stretches of time. There are hundreds of thousands, perhaps millions, of species of beetle. Every single one is built on the same basic plan. They vary in size, colour, sexual ornamentation and much else. But they are all recognisably beetles. The same is true of the other great lineages of animal life. The main division of the kingdom of animals is into *phyla*. There are thirty odd: the exact number is in dispute. Some are scarcely known as fossils at all. But all of those that have decent fossil records appeared early. That leads Gould to the view that the main ways of building an animal were all invented at roughly the same time, and no new ones – no new fundamental body organisations – have been invented since. Evolution has certainly not ground to a halt when it comes to inventing new adaptations. But if Gould is right, it does seem to have ground to a halt in inventing new phyla of animals. Gould sees this as the most striking fact evolutionary theory must explain.

Moreover, Gould has a different conception of the mechanism of evolution. He argues that selection is constrained in important ways by the limits of variation in lineages. For selection can act only to magnify and sculpt variations found in the population. Moreover, he thinks chance has played a pivotal role in the history of life. In times of mass extinction, many species disappear. Surviving, in Gould's view, depends more on luck than on fitness. So in explaining evolutionary history, Gould places less weight on selection than does Dawkins. Moreover, he has a different view of the way selection works. He is very sceptical of gene selection, for he doubts that particular genes usually have a consistent enough effect on the fitness of their bearers for Dawkins' story to make sense. The effect of a particular gene on a body depends on the other genes in that body and on many features of the environment in which the organism develops. So Gould thinks that when selection acts, it typically acts on individual organisms. But this is only part of the story. Gould is sympathetic to theories of 'species selection'. Species themselves may have properties which make them more, or less, apt to go extinct or to *speciate*; that is, to give rise to daughter species. For example, there are very few species of asexual vertebrates; just the odd species of lizard, fish and frog. Moreover, those few seem not to have long evolutionary histories. A mutation is a copying error that takes place when a gene replicates. Most mutations are neutral or bad, but sometimes they cause a beneficial change. And in an asexual species, if two good mutations occur in separate mother–daughter clones, they cannot combine their luck. If they could mate, they could combine their advantages. So perhaps asexual species are vulnerable to extinction as a consequence of their evolutionary inflexibility.

These differences within evolutionary theory are exacerbated by different assessments of science itself. As *Unweaving The Rainbow* shows, Dawkins is a wholehearted son of the Enlightenment. We should embrace the scientific description of ourselves and our world, for it is true (or the nearest approach to truth of which we are capable), beautiful and complete. It leaves nothing out. Gould, on the contrary, does not think that science is complete. The humanities, history and even religion offer insight into the realm of value – of how we ought to live – independent of any possible scientific discovery. And while Gould has never embraced the view that science is just one of many equally valid perspectives on the world, he has often written of social influences on scientific views. Scientific orthodoxy *does* respond to objective evidence about the world, but often slowly, imperfectly, and in ways constrained by the prevailing ideology of the times. In short, Dawkins, but not Gould, thinks of science as a unique standard-bearer of enlightenment and rationality.

Part II
Dawkins' World

2 Genes and Gene Lineages

The Selfish Gene begins with a creation myth. Dawkins asks us to imagine a primitive, pre-biotic world – a world in which physical and biochemical processes make available a soup of chemical and physical resources. In this soup, nothing lives, nothing dies and nothing evolves. But then, Something Happens. A *replicator*, by chance, comes into existence. A replicator is a molecule (or any other structure) that in the right environment acts as a template for its own copying. Active replicators have characteristics which determine their prospects of being copied, though their chances will always depend as well on their environment. A replicator that is highly copy-worthy in one environment might, for example, be too unstable and hence have very poor prospects in a hotter chemical soup, or one composed of different compounds.

The formation of the first active replicator is a world-shaking event. It is truly something new under the sun, for it introduces natural selection and hence evolution into the world. No copying process is perfect. Hence at some stage, after some number of copyings, the copies of the prime replicator will begin to vary from one another. A population of variants comes into existence. Within the population of variant replicators, some will have better prospects than others. Some will have a higher propensity to be copied. Others will have a lower propensity; they are less stable, or require a less common ingredient in the soup. That creates the conditions for natural selection. For resources are not infinite: the replication of one lineage will have consequences for other lineages. And thus evolution driven by selection begins:

> *Competition + variation + replication =*
> *natural selection + evolution.*

The replicators that descend from the original are weeded by natural selection: the variants with features that promote replication will become common; the variants with features that make replication less likely will become rare or extinct.

It would be hard to exaggerate the differences between a world in the first stage of evolution, and our world. Today's genes are made from DNA: specifically, they are sequences of the four bases *adenine*, *guanine*, *cytosine*, and *thymine* (usually abbreviated to A, G, C and

T) attached to a sugar and phosphate spine. Some genes do nothing at all. But most that do anything code for a protein. Indeed, when biologists speak of genes (for instance, when talking of the number of genes carried by particular organisms) they usually have in mind the base sequence that specifies a particular protein. This specification is implemented by an almost universal code. The base sequence is read in groups of three, each of which (aside from a stop signal) specifies one of twenty amino acids. Hence long base sequences specify amino acid sequences, and such sequences are the 'primary structure' of proteins. The process by which genes produce proteins is indirect, requiring two RNA intermediaries – known as messenger and transfer RNA – and it depends on complex cellular machinery.

The upshot is that genes, and the gene-to-protein system, are themselves complex products of evolution. The first replicators were certainly not DNA sequences. They may have been RNA sequences (in which *uracil* replaces thymine) though even that is very controversial. Moreover, this was a world of the 'Naked Replicator'. In our world, the genes are replicated, and the organism interacts with the environment both to protect the genes and to secure the resources for their copying. Hence biologists distinguish the *genotype* of an organism (the complement of genes it carries) and the *phenotype* (its developed form, physiology and behaviour). But in this

DNA, a double spiral of sugar-phosphate backbones.

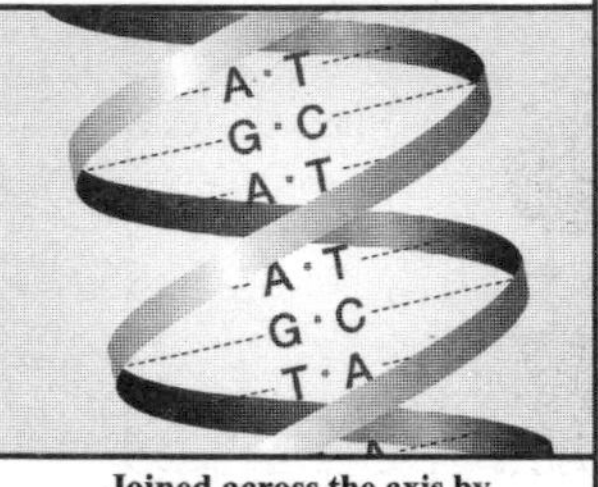

Joined across the axis by complementary base-pairs (A to T and C to G).

Genetic information is encoded in the bases' four-letter language.

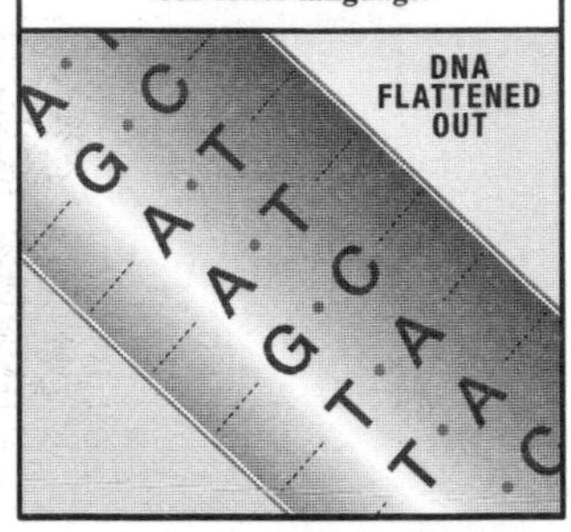

The code on one strand is transcribed by an enzyme causing a complementary molecule to be made.

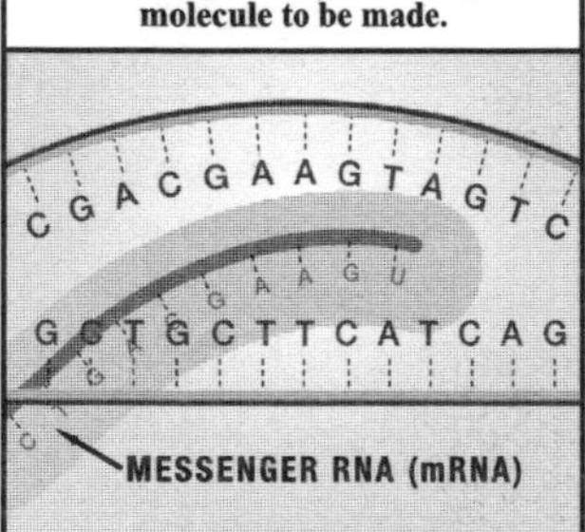

mRNA travels to a ribosome - a double ball of proteins and RNA - to translate its code and to make a protein.

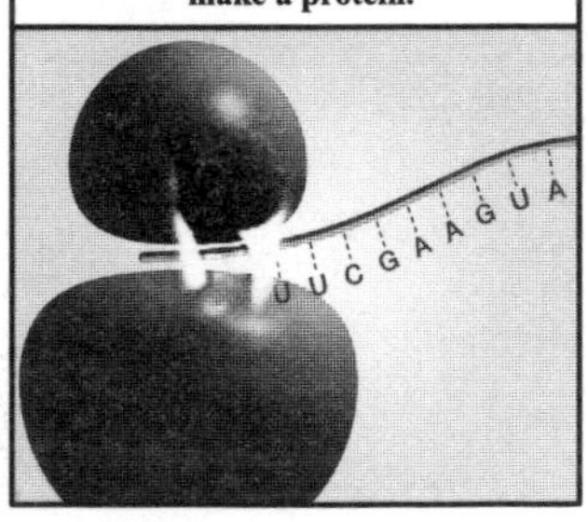

Each triplet of bases on mRNA calls up a complementary triplet on the head of transfer RNA (tRNA).

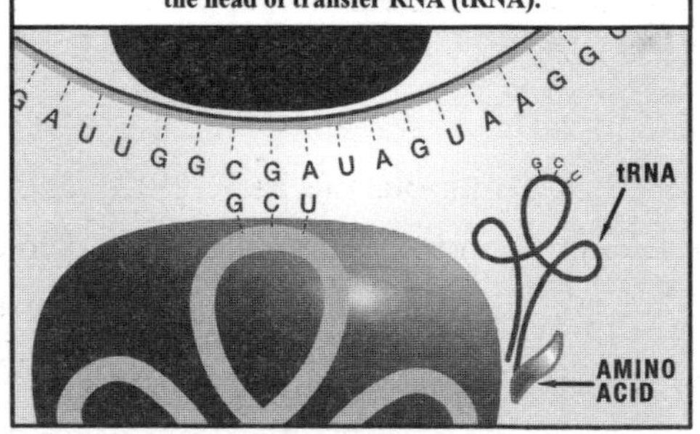

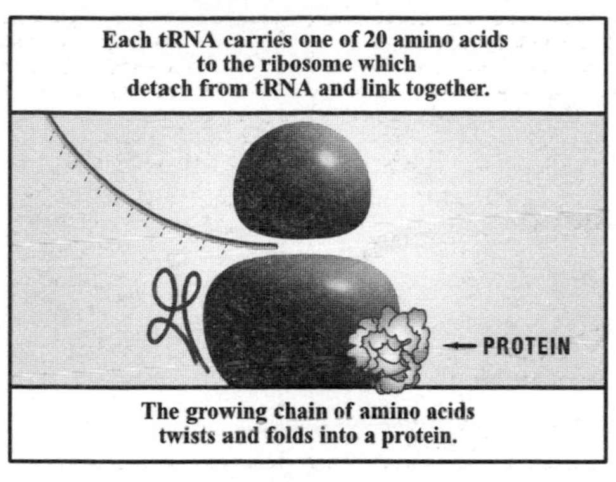

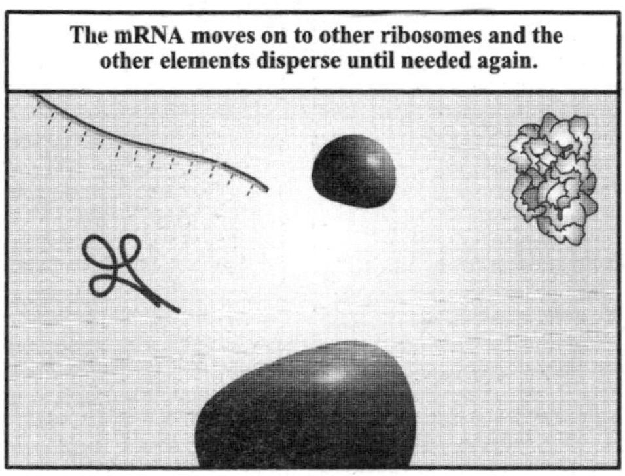

Figure 1: Genes in a DNA sequence code for specific proteins. (Source: *After* Borin Van Loon and Steve Jones, *Introducing Genetics*, Cambridge: Icon Books.)

first phase of evolution, the same entity acts both to secure resources and as a template for its own copying. There is replication and interaction with the environment, but there is no specialisation of roles. In a world in which there are no organisms to 'clothe' the replicators –

no vehicles – selection will not build better-adapted vehicles. Rather, it will select for molecular properties of the replicators themselves. In Dawkins' memorable phrase, we shall see selection for 'fidelity, fecundity and longevity'.

There is some debate about the aptness of Dawkins' creation myth for, in some views, life had its origins in a proto-cellular structure which had no chemicals specialised for the role of replication. But if the first evolutionary regime was as Dawkins depicts it, there is no doubt that in *that* regime, replicators are the units of selection. For nothing else is undergoing evolutionary change. Perhaps these first replicators are not really living things, but certainly nothing else in this world is at all life-like.

Geologically speaking, this first evolutionary regime cannot have lasted for very long. For shortly after the earth became inhabitable, fossils of bacteria-like organisms appear in the record; these are found in 3.5-billion-year-old rocks from the Pilbara region of West Australia. So in the space of a few hundred million years, at most, primitive replicators must have combined into physically and functionally united alliances which formed the first cell-like structures. By that stage, as well, genes built from DNA sequences had probably supplanted the ancient replicators whose origin initiated the whole evolutionary process. Life – for bacteria are uncontroversially alive – had by then passed the organism threshold.

The 'invention' of the organism, perhaps even of the simplest cells, changes the character of both evolution and selection. For the invention of the organism is the invention of a specialised vehicle for protecting replicators, and for harvesting the resources they need to make new copies of themselves. And some vehicles will be better adapted to their circumstances than others. They will succeed *differentially*. This differential ecological success of vehicles causes the differential replication of the replicators that built those vehicles. If very venomous snakes are more ecologically successful than less toxic variants of the same species, the replicators associated with the venomous variants will be replicated more frequently. In the gene pool of that species, gene lineages associated with the venomous variants will replace lineages associated with less toxic snakes.

So after the organism threshold is crossed, natural selection typically acts directly on organisms and indirectly on replicators. Moreover, it selects replicator *teams*, the whole vehicle-building genome, rather than individual replicators. For if an organism dies, all the replicators in it are destroyed. If it succeeds in reproducing, each replicator in it shares in that success; or, at least, has an equal chance of sharing that success. The genes in an organism typically depend on one another;

they share a common fate. Moreover, the effect of a particular replicator on the vehicle carrying it is sensitive to its internal and external environment. Penis-making genes, for example, do not make penises in female vehicles. And they certainly exist in those vehicles. The Y chromosome that male mammals have, and female mammals lack, carries few functional genes. All male mammals inherit from their mother many genes relevant to their male characteristics. Thus a gene's context can make a vast difference to what it does. A toxicity gene in one context may make no difference at all in another. So everyone can agree that in the world of the Naked Replicator, if there ever were such a world, replicators and replicator lineages are the unit of selection. But perhaps the invention of the organism changes the unit of selection.

3 Gene Selection in a World of Organisms

Even after evolution crosses the organism threshold, Dawkins, and before him G.C. Williams, argues that the gene remains the unit of selection. For, as we saw in Chapter 1, selection is cumulative. In *The Blind Watchmaker* and *Climbing Mount Improbable*, Dawkins emphasises the difference between single-step and cumulative selection. In *The Blind Watchmaker*, he points out that if you tried to generate

methinks it is like a weasel

by randomly picking character strings of the right length by repeated trials, you would still be trying at the end of time. The monkey never gets to type even a single sentence of Shakespeare. Cumulative selection transforms the problem. Suppose you make, say, ten random trials and keep the closest, even if it has only a couple of characters in the right place:

qwtxzuifsautysszyaffqyfnm

You then breed from this closest miss, with some error in the copying process. Suppose each daughter string varies from the parent by one character. If so, one of those daughters is likely to have three characters right.

qwtxzuifsaut sszyaffqyfnm

Breed from that one, and so on. Within a manageable number of generations you will reach 'methinks it is like a weasel', though the exact number will depend on luck and copying accuracy. This example is, of course, not a model of *natural* selection; it is more like *artificial* selection. But it does show the great difference in power between single-step selection and cumulative selection. In *Climbing Mount Improbable*, Dawkins reports on various models of the evolution of the vertebrate eye from a patch of photosensitive skin, models that suggest this transformation might take only a few million years.

Adaptive evolution requires cumulative selection. In turn, cumulative selection requires persistence. Proto-eyes must be exposed *repeatedly* to selection's scrutiny, if efficient eyes are to evolve. No such eyes will evolve if the proto-eyes of generation 1000 are significantly different from those of 999 or 1001. Persistence requires copying; neither individual genes nor individual organisms exist for long. Genes are copied into succeeding generations but organisms are not. Hence the unit of selection must be a lineage of gene copies. QED.

Well, not quite. In *The Nature of Selection*, Elliott Sober has pointed out that persistence need not involve copying. For efficient eyes to evolve, the evolving population must be under selection for vision. The eyes of generation N+1 must indeed be similar to those of generation N. So the design of eyes must persist across the generations, to serve as a basis for further improvement. But no eye of generation N+1 is a copy of any eye of generation N. Every eye is built from scratch through a complex developmental process. To be sure, genes play an absolutely crucial role in embryological development, but no eye-copying is going on. DNA acts as a template for its own copying. So, perhaps, do some cellular structures as one cell splits into two. But eyes certainly do *not* act as templates for their own replication. Heritability – the similarity of organisms across generations – is essential for cumulative selection. But the heritability of vision does not involve copying eyes themselves. Perhaps, though, it involves copying the eye-making program. Perhaps something has to be copied, for eye design to persist across the generations.

One popular idea is to suggest that though eyes themselves are not copied, the information needed to make eyes and the rest of the creature is coded in the genes. That information is copied and used. The idea that the genes make up a program or information store is widely endorsed. But it turns out to be surprisingly hard to show that there is a way in which genes and only genes can carry information about a developing organism. Many resources are needed to build an organism. Why think that only some of those – the genes – tell a story about what the organism will be like?

A simpler thought is that though eyes are not copied, eye-making genes are. Here, battle is again joined. For Gould and his allies deny the existence of eye-making genes. Everyone agrees that evolutionary change is associated with genetic change. After World War II, the myxomatosis virus was introduced to Australia to control the rabbit plague that ravaged the country, and while it initially wiped out vast numbers of rabbits, disease resistant rabbits soon evolved. When Australian rabbits evolved their resistance to myxomatosis, the gene pool of the Australian rabbit population changed. Evolutionary changes in a population of organisms are tracked by changes in the gene pool of the evolving population. So it is common ground that evolutionary change in a population is correlated with a change in that population's gene pool. But correlation is not the same as causation. The flight of migrating waders to Siberian breeding grounds *correlates* with the start of the Australasian rugby season. But that flight does not *cause* rugby to begin. Similarly, Gould and others reject the idea that the characteristics of genes causally explain evolutionary changes in a population.

At issue is the relationship between genes and the characteristics of organisms they inhabit. Gould would accept the existence of eye-making genes, and selection for such genes, if the eye-maker invariably caused the organism that carried it to develop an eye of a specific kind. In other words, in Gould's mind, gene selection presupposes something like *genetic determinism*. We have to be a little careful in talking about genetic determinism, for no one has ever imagined a gene could make an eye all by itself. Rather, Gould and his allies think that Dawkins is committed to the idea that there is a stable and simple relation between a particular gene and the characteristics of the organism it is in. So, to revert to our rabbits, if there were a specific rabbit gene that always, or almost always, caused a rabbit to be resistant to myxomatosis, then we could say that that gene is a unit of selection, and that it has replicated vigorously in Australia because it makes rabbits myxomatosis-resistant. Selection sees through the phenotypic trait of the rabbit – disease resistance – to preserve and copy the gene lineage that is responsible for that trait.

Some genes – those that are *invariant* – do have the same effect on an organism in pretty well every circumstance. They can be quite common in bacteria because the development of the bacterial cell is much simpler than the development of any multi-celled organism. A bacterium acquires an appropriate plasmid – a small parcel of genes – from another bacterium and that acquisition suddenly makes the bacterium and all of its descendants resistant to an antibiotic. For bacteria, the whole problem of differentiation – how different cells move to their adult position and specialise – does not arise. But in multi-celled organisms, an invariant relationship between a gene and the organism it travels in is the exception. When such a relationship is found, it is usually bad news. Most genes with invariant effects cause genetic diseases; they are invariant because they cause something to go badly wrong. Even amongst genes that cause genetic diseases, a simple relation between a gene and its effect on the organism is the exception rather than the rule. The more usual situation is that most characteristics of organisms are influenced by more than one gene. Malaria resistance in humans is a case in point: it is a consequence of having both the standard haemoglobin-making gene and a variant form of that same gene; the so-called sickle-cell gene. The effect of any particular gene is typically variable and context dependent. In fact, a person with two copies of the sickle-cell gene is in trouble and will probably die of anaemia. If they had instead just one copy of that gene, and one copy of the normal form, they would be fine.

The upshot is that the relationship between genes and

organisms is typically complex and indirect. There is no simple link between genes and traits. No gene makes a trait; few genes are invariably connected with a specific trait. These basic facts are uncontroversial. However, gene selectionists think that Gould and his allies overstate their importance. Dawkins' views demand that genes have *phenotypic power.* They influence their environment in ways connected to their propensity to be replicated. But that influence can and does depend on their genetic, cellular and ecological environment. So, in *The Extended Phenotype*, Dawkins points out that there are 'genes for reading'. Of course, no gene *causes* its bearer to read, come what may. But a gene has phenotypic power over reading if, if substituted for its rivals for the same slot on human chromosomes, the resulting individual is *more likely* to be able to read. Gene selection requires that degree of consistency in a gene's phenotypic effect. But it requires no more: Dawkins, Williams and other gene selectionists are not committed to genetic determinism or anything like it.

So far then, a stand-off. Cumulative selection is fundamental to evolution. But that alone does not establish that the fundamental agents in evolution are gene lineages, and the sceptics are right to say so. But equally, Dawkins is right to resist the attempt to hang the albatross of genetic determinism around his neck. In the next chapter, we consider Dawkins' attempt to break this deadlock.

4 Extended Phenotypes and Outlaws

Crossing the organism threshold changes the way selection acts on genes. Before that threshold is crossed, and before genes were assembled into co-operating teams, evolution was a war of all against all. By 3.5 billion years ago, that must already have changed. The genes that built 3.5-billion-year-old cynobacteria were not lone wolves. The relations between gene lineages had already become a balance between competition and co-operation. For no one gene can build a vehicle, not even one as relatively simple as a bacterium. So if the advantages of protection, chemical manufacture, and resource harvesting offered by a cell were to be seized, genes had to form alliances of gene complexes. Particular genes in those assemblies were to influence specific characteristics of their vehicles.

The success of one gene lineage has implications for the success or failure of others. Rabbit gene lineages in Australia are in competition with sheep, kangaroo, and wombat gene lineages. There is limited ecological space for grazing animals, and so limited ecological space for their genes. The success of a rabbit-building gene lineage may blight the prospects of gene lineages producing wombat fleas. In a crowded and interconnected world, the triumphs of one lineage will send causal ripples into many parts of the pool. Flea genes and wombat genes, though, are not irrevocably fated to be competitors. Quite often their fates will be independent of one another. Sometimes, over evolutionary time, different gene lineages in different organisms can become allies. Many fungus genes are allies of tree genes, because there are many associations of mutual benefit between fungi and trees. As Bert Hölldobler and Edward O. Wilson show in their *Ants*, such associations are also very common between ants and trees – many ant genes are allies of tree genes. Equally, genes that travel together in the same organisms are natural allies. For they typically succeed or fail together.

There is a case where competition is inescapable. The different *alleles* of a gene are the different DNA sequences within a species that can be found at the same location on a chromosome. These are different versions of the same gene. They are fated to compete with one another, for alternative alleles in a breeding population are rivals for particular slots on the chromosomes of that population. Triumph for one allele means extinction for the others. Australian magpies are co-operative breeding birds that live in extended families. They are famous for defending their nests with vigour, even against people. The magpie breeding season is a time of terror for cyclists and small children. If a particularly aggressive type of magpie founds new families at a greater rate than others, that differential reproduction will cause a differential replication of the gene or genes responsible for that increased aggression. Copies of the aggression gene will rise relative to alternative alleles that compete for the same slots on the magpie chromosomes. The alternative alleles will drop as a fraction of the population, perhaps to zero. Their lineages will become thinner. So the struggle between rival genes is carried out by collectively constructed organisms that mediate both their interaction with the environment and their further replication.

This is the normal pattern of gene action. Both Dawkins and his opponents can tell a reasonable story about this case. Dawkins' story will be about genes and vehicles. Gould, Sober and others will describe the evolution of magpie aggression in terms of the fitness of individual magpies. But this is not the only way genes lever their way into the next generation. Some genes are loners. 'Outlaw genes' promote their own replication at the expense of the other genes in their organism's genome. Outlaws are relatively uncommon but not unknown. One set of examples are 'sex ratio' distorters. In most circumstances, selection on individual organisms favours a sex ratio of 50/50. But not all genes are equally likely to end up in each sex. Most of the genetic material of complex animals like humans is organised into chromosomes. In our normal cells, we have 46 chromosomes, organised into 23 pairs. These cells are *diploid*. Each gene then exists in two versions, one on each chromosome. Each of these is an allele. These versions can be identical, in which case biologists describe the organism as being *homozygous* at that locus. Or they can be different, in which case the organism is *heterozygous*. When the sex cells (the *gametes*) are formed, this number is halved. Each chromosome pair gives rise to just a single chromosome in the sperm or egg, with its genetic material drawn from the two paired parental chromosomes. So our sex cells have 23 chromosomes. They are *haploid* cells, in contrast to normal cells with 23 chromosome pairs. These haploid cells are formed by a special kind of cell division called *meiosis*.

In most cases, when a 23 chromosome cell is made from a 46 chromosome cell, any particular gene in the parental cell has a 50/50 chance of making it to a sperm or an egg. But not in every case. Some genes are only ever passed on to male offspring; others only ever make it to daughters. Mammals like us have a sex determining process that depends on the nature of one of the chromosome pairs. A fertilised egg results when sperm and ovum fuse. In our case, each contributes 23 chromosomes and thus builds a new diploid cell with 23 chromosome pairs. A fertilised egg with two X-chromosomes is female; a fertilised egg with an XY pair is male. So sex determination in mammals depends on the male: all female gametes have is the X chromosome. No gene on the Y chromosome ends up in a female. So there is selection at the gene-level for any mutation on the Y chromosome that biases the sex ratio towards males, even if that mutation reduces the fitness of the organism with it. Males have an XY sex-specifying chromosome pair, so when they produce sperm, they make some X-carrying sperm (daughter-makers) and some Y-carrying (son-making) sperm. So imagine a mutant gene on the Y-chromosome that produced fast sperm, sperm more likely to reach the unfertilised egg first. There would be selection at the genetic level in favour of speedy-Y sperm, even if males were in general less fit, being too numerous to have a good chance of finding a mate. Genes often have more than one effect on their carrier. So the speedy-Y gene might have extra, and unfortunate, effects on the male that carries it. But there can be selection for the speedy-Y gene even if speedy-Y males were less fit than other males.

There can be female-biasing outlaws too. A gene that is copied into all an organism's female offspring and only them, and which made that organism more likely to have female offspring, would have a higher fitness than the other genes. We have genes which are only inherited through the female line. For though most of our genetic material is organised into these 46 chromosomes, not all of it is. *Mitochondria* are energy-generating structures that exist outside the cell nucleus and possess their own genetic material. They are typically inherited *maternally*: your mitochondria are almost always inherited from your mother. For the sperm consists of nothing much more than its genetic warhead and a tail (powered by only a few mitochondria that are later discarded) to propel it, whereas the ovum is a whole cell, stocked with cytoplasm, nucleus and many mitochondria. I am a male, and consequently my mitochondrial genes have no chance of making it to my children. Conversely, my partner's mitochondrial genes are certain to be in any of her children. We have a daughter, but if our daughter had been male, the mitochondrial genes which my partner passed on would be in an evolutionary dead-end. So any mutation in those genes that made her more likely to be a daughter-maker would be favoured. There would be selection in favour of any mitochondrial mutation biasing the sex ratio towards females, even if there were deleterious consequences for individual fitness. Such genes are known in plants. They cause plants that are normally capable of producing both pollen and seed to produce only (mitochondria-carrying) seed. So sex ratio distorters are an example of genes that do not have the same 50/50 chance of replicating as the other genes of an organism.

A second class of outlaw genes are 'segregation distortion' genes. When the gametes are formed in a sexually reproducing organism, the normal complement of chromosomes is halved. Normally each allele on each chromosome has a 50/50 chance of being copied into the gamete. Segregation distorter genes bias this lottery in their favour by chemically sabotaging the allele with which they are paired and thus improving their own chances of making it to the gametes. Hence the segregation distorter gene on one chromosome is fitter than its matched gene, its allele, on its paired chromosome. But in increasing their own fitness, segregation distorters often decrease the fitness of the organism carrying them. For organisms with segregation distorting genes facing each other on matched chromosomes are often sterile.

Outlaws are an uncontroversial case of gene selection. Gould, Sober, Lewontin and many others are sceptical of Dawkins' overall views, but they concede this case to him. But outlaws are not the only case that fits gene selection. In *The Extended Phenotype*, Dawkins argues that there are many examples in which genes reach out into the world to promote their own replication. They have effects, and are visible to selection through these effects. But these are not effects on the organism the gene inhabits. Genes have 'extended phenotypes'.

The most vivid examples of extended phenotypic effects involve the action of parasite genes on host bodies. There are many surreal examples of such gene actions. Parasitic barnacles of the *Rhizocephala* group take over the behaviour of their crab hosts. After attaching themselves to their host, they transform into a single-celled stage that burrows into the crab, grows and then biochemically castrates and feminises the crab (if it's a male) and subverts the host's brood care behaviours so that these are now supportive of the parasite's own eggs. Elliott Sober and David Wilson in *Unto Others* describe a brainworm that burrows into the brain of an ant, changing its behaviour so that it rests on grass leaves waiting to be eaten by a cow. This is not good news for the ant, but it is for the parasite, for the cow is the brainworm's ultimate host. *Wolbachia* is a bacterium that is passed from an infected host to its *female* offspring. It infects various species of insect, and in various ways biases the sex ratio towards females, either by turning its host into a female even if it's genetically male, or by turning its host into an asexual female (i.e. one that reproduces without mating, producing identical female clones of itself).

In all of these cases, alterations in the host are due to

the adaptive effects of parasite genes. A less brutal example of an extended phenotype is house construction by caddis fly larvae. These larvae typically live on the bottom of streams, and glue together an assortment of debris to form a house in which to live. These houses protect the caddis fly larvae in the same way that a clam shell protects its occupant. But the caddis house is not part of a caddis fly body. It is not part of the organism itself.

Parasite manipulation genes and caddis house-making genes do have effects on the bodies in which they travel. There are many links in the causal chain from one replication of a gene to the next, and the chain begins in the parasite's body. The barnacle produces chemical signals that subvert the host's normal behaviours. The subversion gene directs the production of these chemicals. But the adaptive effect of the parasite's gene is its effect on the host's behaviour. Suppose we were to ask: why do these genes exist in the genome of every barnacle in the species? We would answer by describing the feminisation of the crab.

For most genes, the path to the future is through their effects on the organism they help build. If a gene contributes to making that organism particularly well adapted, and if this is the case in most contexts that the gene finds itself in, it will be replicated frequently. If it typically depresses the fitness of its bearer, it will decline in frequency. In this core case, Dawkins' conception of evolution as a struggle between gene lineages, and Gould's view that selection acts on individual organisms, are roughly equivalent. However, though this is the most common case it is not the only case. Genes have two other replication strategies. A few genes are outlaws replicating themselves at the expense of others in the same genome. Their replication-enhancing effects are not effects on the organism in which they ride. Speedy-Y genes do have effects on speedy-Y males: they make it harder for such males to find mates. But that is not the effect which explains why speedy-Y genes would spread in the population. A speedy-Y gene's adaptiveness lies in its local effect on the gamete that carries it. Extended phenotype genes are not outlaws. The barnacle gene that feminises the crab enhances the prospects of *every* gene in the barnacle genome. But it influences the environment of its vehicle, the barnacle, rather than the barnacle itself. Its adaptive effect is outside the body it inhabits. Outlaw genes and extended phenotype genes cannot readily be incorporated within a view of evolution that sees selection as acting on individual organisms. Here Dawkins' view of evolution seems to be doing better than that of Gould. A special case of extended phenotype effects concerns social behaviour; the behaviour of groups of animals. To this, we now turn.

5 Selfishness and Selection

Dawkins' most famous book, *The Selfish Gene*, is a response to a pressing evolutionary problem. How could co-operation have evolved? Co-operation is common in the animal kingdom. Many animals co-operate in defending themselves against predators. Musk oxen physically defend themselves as a collective. Many species of jay and crow collectively 'mob' hawks, owls and other dangerous birds. Less spectacularly, but more commonly, many animals warn one another of danger with distinctive calls. A number of carnivores – wolves, African hunting dogs, chimps, lions and at least one species of hawk – hunt co-operatively and share their kills. Lionesses tolerate suckling by the cubs of their pride mates. Vampire bats that have failed to find blood successfully beg from other bats in their roost. Many species of birds breed co-operatively – parents have 'helpers at the nest', contributing to both the defence of the nest and to feeding the nestlings. White-winged choughs, for example, can only breed with helpers. A single mated pair has no chance of successfully raising fledglings.

So co-operation is far from rare. Yet it poses a puzzle familiar also from human society. For co-operation seems to be *altruistic*. It is true that *everyone* is better off if *everyone* co-operates. Everyone in the tribe is safer if everyone fights bravely in its defence. But I am better off still if I quietly withdraw to safety while *everyone else* fights bravely. This is known as the 'temptation to defect'. It has an evolutionary parallel. Think, for example, of a vervet monkey that has just noticed an eagle. Wouldn't it be best off just quietly hiding? Calling can attract the eagle's unwelcome attention. Over time, we would expect selection to cull such traits as warning others about predators, signalling the presence of food, contributing to collective defence, and caring for others' young.

What then could explain altruism? There seem to be three possibilities. Altruism might be inadvertent. Animals are not perfectly adapted to their environment. For example, no recognition system is ever perfect. Some risk of error is inevitable. So perhaps the lioness tolerates another's cub at her nipple rather than risking the rejection of her own cub by mistake. Tolerating an occasional freeloader would be much less expensive than rejecting her own young. So if there is a chance of such a mistake, caution is reasonable. Yet that caution makes her vulnerable to freeloaders. This possibility might explain a few examples, but it's hard to see how an animal could engage in collective defence or warn others of predators by mistake. No error hypothesis can explain all instances of altruism.

A second idea is to explain altruism as a result of selection on the collectives of which they are members. In some species of baboon, the adult males defend the troops of which they are a part. In this view, the collective is itself a vehicle of selection; it is a 'super-organism'. The co-operative baboon troop is more likely to survive and is more likely to found new troops like itself than one composed of baboons following the maxim of 'every baboon for itself'. According to this suggestion, the population of baboon troops is a salient level of biological organisation. It forms a population of mini-populations (the troops) where these mini-populations compete with differential success.

A third possibility is that altruism is an illusion. The idea here is to explain away the appearance of altruism. This possibility is central to contemporary debates on this issue. Berent Heinrich's *Ravens in Winter* explores one such case. Heinrich was puzzled by the fact that when an individual raven found a carcass – a rich source of food – it seemed to advertise its find rather than attempt to monopolise it. Why would a raven do such a thing? It turned out that ravens who call at the sight of large carcasses are not acting altruistically at all. Adult ravens hold territories, whereas juveniles do not. The ravens that call when they discover a carcass have no territories of their own. By themselves they will be chased off by the territory owner and end up with nothing or almost nothing. So they call and in doing so they recruit others. The recruits swamp the owner's territory defence. The calling ravens will have to share with those they recruit, but they will still get some of the windfall.

In pursuing this line of thought, evolutionary biologists have been particularly interested in two ideas. One is the idea that co-operation involves trading benefits. If two or more animals can secure some resource by co-operating that neither could secure individually, individual selection could promote joint action. Social predators – such as wolves and African wild dogs – take and then share prey that no individual could kill by itself. It is in each dog's interest to act with the others, so long as the individual's share of the joint carcass is more valuable than any prey it could catch itself. Reciprocal altruism takes this unproblematic form of co-operation as its model and extends it to cases in which the partners do not reap their reward simultaneously. Each is better off trading than not trading, and each animal in the trade is vigilant to ensure it is not cheated. This view of co-operation has been strengthened in the last decade and a half by the work of Robert Axelrod, who has shown that the strategy of 'tit-for-tat' can pay off in many situations. Tit-for-tat is the rule of co-operating on the first interaction with another animal, and thereafter doing what it did last time. So if your partner defected on the first interaction, i.e. if it failed to co-operate, on the second interaction you defect back. If your partner co-operated, you co-operate. The best known biological example is from the life of vampire bats. They share blood. These bats die unless they feed every couple of days, and hunting failure is quite common. So reciprocation is an essential element in vampire bat life. Successful bats share with those who fail. But bats that give are bats that receive.

The problem of social interaction, especially the problem of co-operation, has driven the development of many of the most important new ideas in evolution. One particularly important idea is evolutionary games theory. When an animal interacts with its environment, its fitness does not usually depend on the other members of its population. A gene in a tiger that improves its visual acuity or metabolic efficiency will benefit that tiger, whatever other tigers do. Such traits are beneficial independently of their frequency in the population. But the fitness effects of social traits often depend on their frequency. Even if wolves would be better off hunting co-operatively than going it alone, a wolf with the disposition to co-operate will gain no benefit thereby unless others have it too. Other traits have the reverse dynamic. In a population of co-operators, the rare cheat does very well. John Maynard Smith developed a famous model to show that social traits, including co-operative ones, often have equilibrium frequencies at which two opposed traits both persist in the population.

Maynard Smith imagined a population that had no real contests over important resources; for example, a nesting hole. If two birds both wanted the same hole, both would bluff for a while, and then one would give up. No bird would actually fight for a hole. These birds follow a 'dove' strategy in their interactions. This population would be vulnerable to an invasion by a bird playing 'hawk', actually attacking to fight for a hole. In a population playing 'dove', hawks do superbly. They always get their hole, and they never have to pay the cost of actually fighting. But as the frequency of hawks rises in the population, the cost of being a hawk rises too. For now they begin to meet other hawks, not doves, at the holes. They do not always get their hole, and they have to bear the costs of fighting. Maynard Smith showed that unless holes are very valuable, or the risks of fighting are low, there will be an equilibrium frequency of both hawks and doves in the population (or a frequency with which each bird sometimes acts like a hawk, and sometimes a dove). This frequency is evolutionarily stable. Axelrod's work shows that, in important circumstances, tit-for-tat is an evolutionarily stable strategy. A population all following this strategy cannot be invaded by any mutant following an alternative strategy.

A second strategy for explaining away altruism is based on another unproblematic form of co-operation. Many animals aid their own offspring. In doing so, they project their own genes into the future, for their young bear their genes. But that is true not just of an animal's direct offspring. An animal's kin, especially its close kin, carries copies of its genes. Sometimes an animal can best project its genes into the future by aiding kin. Behaviours that evolve through benefit to kin are known as kin-selected effects. The measure of fitness which includes these indirect effects is known as 'inclusive fitness'.

Gene selection was born when G.C. Williams, in *Adaptation and Natural Selection*, defended the idea that altruism is an illusion against group selection theories of altruism. Williams argued that selection at the level of groups would in almost every case be undermined by individual selection for defection pushing in the opposite direction. The evolutionary temptation to defect is too great for real altruism to evolve. A selfish individual in an altruistic group would always be fitter than his fellows – so the altruism of the group would be eroded from within. Moreover, wolf packs and similar groups last longer than the individuals within them. The group's life span is longer than the individual life spans. So the clock of *individual* selection runs faster and hence more powerfully than that of *group* selection. In this respect, *The Selfish Gene* follows in Williams' footsteps. For Williams and Dawkins, co-operation is real but altruism is not. So they aimed to explain away the appearance of altruism. They are both very sceptical about group selection; of the idea that groups are 'super-organisms'. Groups are not themselves adapted. Rather, they are shifting ensembles of individual organisms. A fast herd of horses is just a herd of fast horses: the adaptation – speed – is a trait of the individual horses in the herd, not of the herd itself. Yet group selection presupposes that the herd itself is adapted.

On this issue the views of Dawkins and Gould have converged to some extent. For one thing, since *Adaptation and Natural Selection* and *The Selfish Gene* were written, it has become clear that gene selection is consistent with group selection. Dawkins agrees that organisms play a central role in evolution. They are vehicles of selection: their success determines the proliferation of gene lineages. The most prominent current defenders of group selection, David Wilson and Elliott Sober, point out that group selection is a claim about vehicles. The group selection theorist can agree that at its most fundamental level, evolutionary history is the history of the success and failure of competing gene lineages. But they claim that some rival gene lineages compete by coding for the characteristics of groups. If groups differ in their ecological success, that difference impacts on the replication of genes carried in those groups. If there are baboon genes which induce baboon groups to defend the whole group aggressively against threats from leopards, that is just another example of a gene with an extended phenotype.

So, in embracing gene selection, Dawkins need not reject group selection. This much is now uncontroversial. Moreover, it is clear that the defection problem cannot be an absolute bar to the evolution of co-operation. For the evolution of the organism itself involves just that problem. The evolution of gene teams, joint phenotype building, joint and fair replication, and sex cell formation all posed co-operation and defection problems. The fitness of every replicator is boosted if all co-operate, say, in building a cell before replicating. But individual replicators would have had evolutionary temptations to defect, to become outlaws. The existence of segregation distorters and other outlaws shows that the defection problem is not fully solved. Cancers are cells that have become outlaws. But the fact that organisms have evolved shows that it can be partially solved.

Gene selection is not intrinsically inconsistent with high-level selection, for defenders of gene selection can accept that groups are vehicles. Moreover, the type of high-level selection Gould defends largely avoids the defection problem. Gould does think that some units of selection are themselves composed of individual organisms. But Gould has in mind *species selection*, not group selection. While showing due caution, Gould is rather persuaded by the idea that species differ both in characteristics that make them vulnerable to extinction, and in characteristics that make them evolutionarily fecund. For example, species whose gene pools contain plenty of variability are, other things being equal, more resilient in the face of environmental change than species with relatively little variability. The same is true of species with broad geographic ranges. Broad-ranging species are more buffered against change and hence less likely to go extinct than those with narrow habitat tolerances.

The distinction between group selection and species selection is important. Group selection and selection on individual organisms are mechanisms that are sensitive to traits of the same kind: warning calls, food sharing, joint defence and the like. That is why selective forces can act in the opposite direction. The fact that group selection favours, say, collective defence and that individual selection selects against joint defence opens the door to the problem of defection and hence to the likelihood that individual selection will be more powerful than group selcction. Only under special conditions can group selection drive an evolutionary change in the teeth of individual-level selection against that change.

This problem does not arise for species selection. For the traits relevant to species selection are not traits of individual organisms at all. Consider the candidates. They include such characteristics as geographic range, gene pool heterogeneity and the like. These are properties of populations not individuals. So the main problem Williams and Dawkins have urged against group selection, the defection problem, does not arise for the version of high-level selection Gould has been exploring. Moreover, Dawkins' own views have edged towards that of Gould. In *Climbing Mount Improbable* (in Chapter 7), he discusses the evolution of evolvability itself. Some lineages of animals are more 'evolvable' than others – something about the basic organisation of the animal makes it easier to generate evolutionary change. In this connection, Dawkins discusses the evolution of body segmentation. Arthropods are animals with exoskeletons and jointed, segmented bodies. Spiders, crabs, and insects are all arthropods. Perhaps it is no accident that the segmented-bodied arthropods are far and away the most diverse lineage of animals. For once segmentation has been invented, natural selection can specialise segments to new roles. Thus their limbs have often been converted into feelers and various other specialised biological machines. Dawkins suspects that the rich radiation of segmented animals is explained by some kind of high-level selection for evolvability, albeit one that is not opposed by individual-level selection.

Sober and Wilson have reopened the debate about group selection, arguing that animals are not just co-operative, they are altruistic. Moreover, there remain many important debatable issues about species selection. But on these issues, the divisions between Dawkins and Gould are less sharp than they once were.

6 Selection and Adaptation

I have just suggested that the fissure lines between Dawkins and Gould on high-level selection are not as wide as they once were. Despite the heat of some recent rhetoric, the same is true of the role of selection in generating evolutionary change. In 1987, Gould collaborated with Richard Lewontin in a famous critique of evolutionary biology, arguing that, as it was then practised, biology was 'adaptationist'. It was not so very obvious what adaptationism was. But two aspects of this sin were clear. Evolutionary biologists were too ready to assume that characteristics of an organism were shaped by natural selection for some function. And they were too easily satisfied that they had discovered that function.

Most evolutionary biologists would agree that the 1987 paper had a salutary effect, stimulating the development of new ways to test evolutionary and selective hypotheses. One method has been to turn a hypothesis into a formal mathematical model, one making quantitative and measurable predictions about a population. Wasps, bees and other social insects have an unusual genetic system: males develop from unfertilised eggs, and, like our sex cells, they have only one set of chromosomes. Queens and workers are female, and develop from fertilised eggs. It follows that if the queen has mated only once, the sister workers in a nest are more closely related to one another than they are to their mother. As with us, they have one chance in two of carrying any particular one of their mother's genes. But they share, on average, three out of four genes with their sister. For they all get the same set of genes from their father, since he has only one set to give. His sperm are all alike. So the sisters share all their paternal genes, and on average half their maternal genes, making three out of four in total. These facts lead to different expectations about the colony sex ratio if it is queen-controlled than if it is worker-controlled. The queen wants more sons than do the workers. So formal models can then be constructed from and compared with actual data to test for worker-versus-queen control.

A second approach has been to develop explicit comparative methods: methods that compare the species under study with its relatives. The idea is to distinguish traits which are adaptations to current circumstances from traits which are inherited from the species' ancestors, by looking at its relatives. Suppose we wondered why the golden-shouldered parrot lays its eggs in holes in termite mounds. The fact that this parrot lays its eggs in a hole, rather than in a nest that it builds, is probably not an adaptation to its specific circumstances. For all parrots lay eggs in holes. So nesting in a hole is probably a trait that the golden-shouldered parrot inherited from its ancestor species. But most parrots lay eggs in natural holes in trees rather than excavating holes in ant nests. Only the golden-shouldered parrot and a few close relatives, all living in rather treeless grasslands, use termite mounds. So this probably is an adaptation to their specific ecological circumstances. Both these techniques are still being developed, but there is no doubt that evolutionary biologists have responded to the Gould–Lewontin challenge.

Even so, whatever adaptationism is, Gould thinks that it is still alive. In his notorious review of Dennett, Gould accuses him of representing an 'ultra-Darwinian' strand of evolutionary thinking; of believing that just about every characteristic of every organism is shaped by natural selection. There are important differences between Dennett and Dawkins on the one hand and Gould on the other on the role of selection in driving evolutionary change. But this is not one of them. Everyone accepts that many characteristics of organisms are not the direct result of selection. Consider, for example, the male king parrot, which is a brilliant red. It is probably red as a direct result of sexual selection: females prefer red males. But nothing like that is true of blood's colour. It is a by-product of selection for blood's real function, that of carrying oxygen to the tissues. Such examples can be endlessly multiplied. Some traits of organisms result from chance fixations of neutral characteristics; indeed, that seems very likely to be true of many of our genetic characteristics. Some characteristics are the result of an inheritance from distant ancestors that is now entrenched in the way an organism develops. As Gould points out in one of his engaging essays, that is very likely why we have five fingers and toes. Some traits are adaptive vestiges. Many cave-dwelling species have non-functional eyes that are vestiges of the functioning eyes of their sighted ancestors. Some characteristics of organisms are by-products of selection for some other characteristic. Human female reproductive anatomy is jerry-built as a consequence of our adaptation for bipedalism.

None of these general truths are at all controversial, though their application to particular cases may be. Nor is there disagreement between Gould and Dawkins on core cases. For example, Dawkins begins *The Blind Watchmaker* with a discussion of bat echolocation. The fossil record of bats is not particularly rich, but even so, no one doubts that echolocation evolved in bats as a means for them to locate themselves in space and to locate their prey. The echolocation system is complex and integrated. It powers a very distinctive type of behaviour that is central to the life history of bats, and it supports only that type of behaviour. These facts allow us to both identify echolocation as an adaptation and to identify what echolocation is for.

Everyone agrees that bat echolocation is an adaptation. As Gould says, 'eyes are for seeing and feet are for moving' (*New York Review of Books*, 12 June 1997). But once we move away from these uncontroversial cases, identifying adaptation is difficult and controversial. This fact is of particular interest to Gould because the application of evolutionary theory to human behaviour has often involved a shift from these clear and unmistakable cases. For instance, there have been claims that sexual jealousy and rape are adaptations. But neither show the adaptive complexity that so uncontroversially marks echolocation as an adaptation. What would show, say, that jealous rage is an adaptation rather than an unfortunate side-effect of our emotional repertoire? Adaptationist claims of this ilk, perhaps rightly, make Gould's blood boil. But here he has no direct argument with Dawkins. Dawkins, unlike Gould, is confident that some human behaviour patterns are adaptations; a confidence he makes clear in an assertive review of Steven Rose, Leon J. Kamin, and Richard Lewontin's *Not In Our Genes*. But *Climbing Mount Improbable* and

The Blind Watchmaker focus on the uncontroversial cases. For Dawkins takes the explanation of adaptive complexity to be the central task of evolutionary biology. He does so precisely because such systems are highly improbable, and hence can be explained only by natural selection. These books focus on clear cases. On these, he and Gould are in agreement.

There are, however, important disagreements. One is over the relative role of selection and variation. Selection acts only on variation generated in a lineage. The developmental biology of a lineage determines the range of variation. That developmental biology is the result of that lineage's evolutionary history. So the variation that is available to selection in a lineage is determined by its history: its history *constrains* its future evolutionary possibilities. Perhaps a chimpanzee with a powerful and prehensile tail would be fitter than chimps as we find them. For it would be well suited both to life in the trees and on the ground, harvesting the best of both worlds. Even so, if no tailed variants were thrown up in ancestral chimp populations, selection cannot make such a chimp. The evolutionary trajectory of a population is hostage both to selection and to the supply of variation.

One major debate in evolutionary biology is about the relative role of the supply of variation and selection in explaining evolutionary change. We can explore this debate through one of Dawkins' own examples in

Climbing Mount Improbable: the 'Museum of All Possible Shells'. Minor details aside, it turns out that shells vary in just three ways: the rates at which they uncoil in one plane (their 'flare'), the rate at which they rise above that plane (their 'spire'), and the rate at which their tube expands (their 'verm'). This makes it possible to represent the space of all possible shell shapes as a cube, where each dimension of that cube corresponds to one of the three ways shells vary from one another. Thus any point in the cube represents a possible shell: a shell made by expanding in a plane at a given rate; expanding its tube at a specific rate; and rising above the plane at a specific rate. Most of these possible shells do not exist, and to the best of our knowledge never have existed. A large chunk of that cube is unoccupied. What explains the missing shells? Are these missing variants impossible for shell lineages to generate? Have shell-building lineages inherited insufficient variation to build these missing shells? Perhaps, instead, they have been edited out by selection; they are too expensive to build, or too unwieldy, or too fragile.

These questions remain unanswered: both about shells in particular and about all the many apparently possible plants and animals that have never existed. Why are there no centaurs? Perhaps they would be too expensive to run, or too subject to back pain. But perhaps six-limbed mammals have simply never been available for

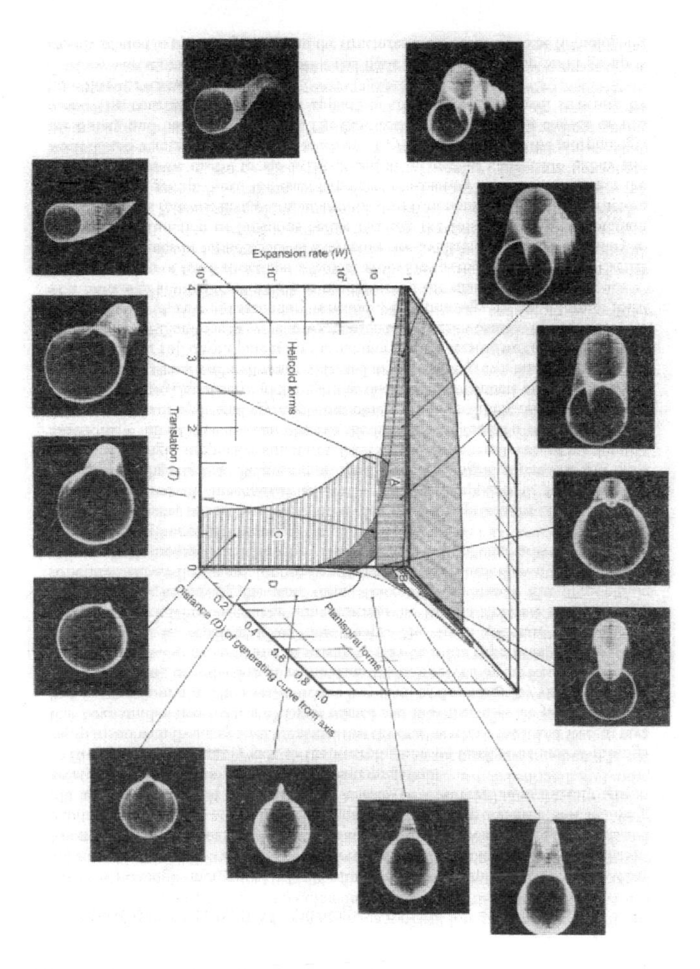

Figure 2: Raup's cube representing the space of all possible shell shapes. Regions of the cube in which real-life shells can be found are shaded. The unshaded regions house theoretically conceivable shells that do not actually exist. (Source: David M. Raup, in Raup and Stanley's *Principles of Paleontology*, London: W.H. Freeman, 1979.)

selection. Dawkins is inclined to make a selectionist bet on these issues. His guess is that, over the long run, the space of evolutionary possibilities open to a lineage is rich. Hence the history of the lineage is largely determined by selection making some of these possibilities actual. Selection determines, for instance, that actual mussel shells are strong, thick and low. Gould, on the other hand, is inclined to bet that the array of possibilities open to a lineage is tightly restricted, often to minor variants of its current state. Hence its history is largely shaped by the events that set the envelope of possibilities; for instance, the events that determined that vertebrates have at most four limbs.

To this difference is added another. For Dawkins, the central problem of evolutionary biology is the explanation of adaptive complexity. That is not Gould's conception of the field. He has spent a large fraction of his palaeontological career arguing for the existence of large-scale patterns in the history of life, patterns not explained by natural selection. So a further disagreement concerns the existence and importance of these patterns. To this, we now turn.

Part III

The View from Harvard

7 Local Process, Global Change?

Gould sees himself as having two fundamental disagreements over selection with Dawkins and others of a similar mindset. One is about evolutionary changes within a species. Microevolution is the branch of evolutionary biology concerned with evolutionary changes taking place within a species; changes that occur at a scale we can observe. Gould thinks that evolutionary biologists too often neglect non-selective possibilities when they formulate and test their hypotheses about microevolutionary change. In his critique of human sociobiology, this has probably been his main beef. For example, E.O. Wilson has argued that males and females differ predictably both in their sexual behaviour and in their behaviour towards children. Men are more apt to be promiscuous than women and are also less inclined to pour all their resources into a single monogamous partnership. Gould is sceptical even of these sociological claims about the way we actually behave, and with reason: Sarah Hrdy in *Mother Nature* has recently given a much more nuanced account of male and female reproductive roles. But Gould is even more sceptical about the adaptationist explanations offered for these activities. Suppose Wilson is right, and that, as a rule, men and women tend to differ in these ways. These differences may be no adaptation at all. Instead, they might be a vestige of sex differences inherited from our hominid ancestors. One of the most famous of all hominid fossils is 'Lucy'; an extraordinarily complete fossil of a female *Australopithecus afarensis*. This hominid species lived about 3 million years ago and was highly sexually dimorphic; that is, the males were much larger and more robust than the females. Perhaps the differences between men and women are just a reduced residue of this much greater original difference. Gould thinks sociobiology has systematically neglected these non-adaptive possibilities, an argument he presses with great vigour in the second of his *New York Review of Books* contributions.

Despite his heat on these issues, this has been the lesser of Gould's concerns. His main target is a view he calls 'extrapolationism'. Extrapolationism concerns the relationship between evolutionary processes that take place within a species and those of the large-scale history of life. Most species are fragmented into local populations, living in environments that vary to some extent from one another. In some species, this fragmentation and isolation can be quite extreme. Barn owls live on every continent but Antarctica, and are scattered through environments that differ in climate, vegetation, competition and predation. They are close to one end of a continuum that runs from very widely distributed and unspecialised species like this owl, to species which live only in one tiny corner of the world. But few species consist of a single homogenous population. Sometimes the local populations into which species are divided are fully isolated; they are cut off in 'islands' of suitable habitat. But, for the most part, there is some migration in and out. Even so, the members of these local populations mostly interact with other locals, both in competition and reproduction. So natural selection takes place within these fragments. Since the different populations are different samples of the variability of the whole species, and because the environments vary, different populations of the same species often diverge from one another, though that divergence is often temporary. It breaks down when populations rejoin.

How do events at this scale relate to the large-scale history of life? How do changes in local populations over a few generations relate to the evolution of species and lineages of species documented in the fossil record? Gould argues that mainstream evolutionary biology has accepted an extrapolationist view. Indeed, he thinks this view dates back to Darwin himself. In this view, the evolution of species lineages is nothing more than an aggregation of events at the scale of local populations. Major changes are minor changes added up over many generations. Evolutionary patterns are generated only by the processes documented in local populations. It is not much of an exaggeration to say that Gould's professional life has been one long campaign against this idea. To begin with, I will review four highlights:

(i) Gould's first famous contribution to evolutionary thinking came in 1972. With Niles Eldredge, he developed the theory of 'punctuated equilibrium', a view of the typical life history of species. According to Gould, species do not gradually evolve into new species. *Homo habilis* did not gradually, imperceptibly turn into *Homo erectus*. Rather, new species typically arise by a split in a parental species followed by rapid speciation of one or both of the fragments. The typical life history of a species involves its geologically instantaneous formation. New species typically appear in the fossil record already fully differentiated from their parent species. Their distinctive characteristics are already present in the earliest fossils, rather than gradually emerging over the species' life history. Once a new species appears, it usually undergoes no further evolutionary change until it goes extinct, or until it splits into daughter species.

Gould argues that the punctuated equilibrium pattern challenges extrapolationism. Extrapolationists ought to expect gradual change in a species. Extrapolationism predicts the gradual evolutionary accommodation of a species, slowly changing itself so that it comes to suit its new environment (for that is how we see local populations respond). Moreover, if species life histories do have this pattern of rapid formation followed by stability, we need a new explanation of evolutionary trends. Hominid evolution is a classic example of an evolutionary trend: over hominid history, there has been a marked increase in relative brain size. But if such hominid species as *Homo habilis* or *Homo erectus* showed no significant evolutionary change after they originated, this trend cannot be produced by a slow growth of relative brain size over the lifetime of a species. Trends must be the result, Gould concludes, of species sorting. Species with relatively larger brains must have been more likely to appear, or to survive.

(ii) In many of his *Natural History* writings, Gould has argued that mass extinctions have had a profound effect on the history of life. Gould was an early supporter of the idea that an asteroid impact caused the Cretaceous/Tertiary extinction, the one that saw the end of the pterosaurs, large marine reptiles and the non-avian dinosaurs 65 million years ago. If a large impact caused those extinctions, then they were sudden, even in ecological terms. Moreover, as Gould reads the record, the big rock did not just finish off a doomed lineage. Had it missed the earth, the dinosaurs might still be dominating terrestrial ecosystems, whales would never have had a chance and mammals might still be rat-sized insect-eaters skulking in the dark.

Mass extinctions do not strike at random. Some types of species are more vulnerable than others. But the level of adaptation of a species is irrelevant. For adaptation is adaptation to a specific environment. Mass extinctions are caused by events which disrupt those environments catastrophically. They suddenly change the rules of the game. Since those changes are sudden and severe, selection is powerless to adapt organisms to their changed circumstances. Dinosaurs were very likely superbly adapted to their habitats, but that is irrelevant if those habitats were destroyed. The properties that are visible to selection and evolution in local populations are irrelevant to the prospects for survival in mass extinction times. Yet survival or extinction in mass extinction episodes determines the large-scale shape of the tree of life. The death of the mammal-like reptiles at the end of the Permian gave dinosaurs their chance; the death of the dinosaurs opened the door for the radiation of mammals.

(iii) In his *Wonderful Life*, Gould describes an extraordinary fauna from early in the history of animal life. The 'Burgess Shale' fauna is known in unexpected detail because a fortunate chance preserved soft tissues, not just the hard parts, of a large number of its members. Gould argues that this fauna demonstrates a completely unexpected pattern in the large-scale history of life. To make his point, he distinguishes between *diversity* and *disparity*. Life's diversity is the number of species in existence at that time. Gould accepts that life over the last few million years is probably more diverse than it has ever been. The disparity of life is measured by the number of basic organisations or body plans that exist at that time. The great richness of beetle fauna contributes hugely to life's diversity but not to its disparity. For beetles are all built in the same general pattern, despite their variations in size, colour and sexual ornamentation.

Having made this distinction between disparity and diversity, Gould makes three bold claims about the overall history of animal life. First, the disparity of animal life was at its peak shortly after multi-celled animals came into existence in the Cambrian, about 530 million years ago, and has since shrunk. Second, very little disparity has been generated since the Cambrian. Those lineages that have survived have shown profound evolutionary conservatism. Contemporary disparity largely consists in the surviving elements of that initial burst of evolutionary experimentation. Third, Gould argues that survival has been 'contingent'; if we replayed the tape of history from the earliest Cambrian, with small alterations in the initial conditions, we might well have a very different set of survivors.

(iv) In *The Spread of Excellence*, Gould argues that evolutionary trends are not the scaled-up consequences of competitive interactions amongst organisms. One much repeated example of an evolutionary trend is the evolution of horses. Horse evolution saw a switch, the story runs, from browsing on trees and bushes to grazing on grasses. This generated correlated morphological changes. Horses became larger, with longer, higher-crowned teeth, and at the same time lost their toes as their feet turned into hooves. If it were the right view of horse evolution, the example would be a triumph of extrapolationism. The evolutionary pattern in the horse lineage would be the aggregation, the summary, of a multitude of interactions in particular populations, most of which had the same upshot. But Gould reinterprets this history. It is not a result of the competitive success of grazers with those characteristics over browsing horses and less well-designed grazing horses. Rather, Gould argues that this trend is really a change in spread of variation within the horse lineage. The horse lineage used to be species-rich, with a wide range of horse lifestyles and sizes. But only a very few species survived, and those few happen to be largish horses. The average horse is larger now only because almost all horse species went extinct, and the few survivors happened to be somewhat atypical.

In considering Gould's case against extrapolationism, two issues will loom large. Are the patterns in life's history that he claims to detect real? And do these patterns really show the existence of evolutionary mechanisms other than those operating at the scale of local populations?

8 Punctuated Equilibrium

In 1972, Gould and Eldredge argued that the fossil records of most species show no significant change from the time they are first identified to the time they disappear. The fossil record is incomplete. Many species are known from only a few strata, though the record of marine invertebrates – and Eldredge's own special group is the *trilobites* – tends to be somewhat more complete. Moreover, with a few spectacular and famous exceptions, only the hard parts of animals fossilise, such as shells, bones and teeth. So some changes would be undetectable. Gaps in the fossil record make evolutionary change seem jerkier than it really was, for intermediates that did exist are not found as fossils. Even so, Gould and Eldredge argued that the appearance of stability was not a mere effect of the gappiness and imperfection of the fossil record. In most cases, this appearance of evolutionary stasis reflects reality. Most species come into existence relatively rapidly, having acquired their distinctive characteristics, and do not significantly change thereafter. By 'rapidly', Gould and Eldredge mean rapidly *by geologists' standards*. Species often exist for a few million years, and the resolution of the fossil record is coarse. In most circumstances, a speciation process that took 50,000 years would seem instantaneous. Yet that is only 2.5% of the 2 million years, say, that the species existed. So a species that took that little time to appear, but which then persisted without significant change, would certainly conform to the punctuated equilibrium pattern.

This hypothesis has been misunderstood in two important ways. In some early discussions of the idea, the contrast between geological and ecological time was blurred. Hence Gould and Eldredge were interpreted as making a very radical claim: species originate more or less overnight, in a single step, with all their new structures present. But that was a misreading. Occasionally plant species do arise in this way by hybridisation between parents of different species. But it is certainly unusual for animal species to originate in a single generation. Gould and Eldredge agree that new structures are almost always assembled over a number of generations, rather than all at once by a macromutation. Speciation – the sundering of a single lineage into two – takes generations.

In recent work, they have clarified a second misunderstanding. In claiming that species typically undergo no further evolutionary change once speciation is complete, they are not claiming that there is no change at all between generation N and generation N+1. Lineages do change. But the change between generations does not *accumulate*. Instead, over time, the species wobbles around its phenotypic mean. Jonathan Weiner's *The Beak of the Finch* describes this very process. In wet years, there is selection for slender beaks that enable finches to eat small soft seeds. In dry years, there is selection for more robust beaks. These are suited for cracking the larger harder seeds available in droughts. Wet years are interleaved with dry ones, so there is no long-term directional selection. The mean size and shape of the finch beak wobbles to and fro. If this fluctuating environment persists over the long term, finch species will be in stasis, as Gould and Eldredge define it. There will be no long-term shift in finch phenotypes.

How frequently do species stay in stasis over their life spans? This issue is still open, but let's grant that stasis is common. Why suppose this is bad news for extrapolationist orthodoxy? Gould and Eldredge agree that new structures are created by cumulative selection over many generations. No departure from orthodoxy here. Moreover, the example of the Galapagos finches shows that we can explain stasis by extrapolation from processes we can observe in local populations. If we aggregated the data from many Galapagos seasons, the result would be a 'wobble' around an average finch beak. There are other local processes that generate stasis, too. For organisms can track their preferred habitats as the environment changes, rather than staying put and adapting *in situ*. The fossil record of the Pleistocene shows that the geographic distribution of many animals shifted in response to climate changes, and we can see similar changes in range on human time scales. Many areas of Australia have been converted from woodland to grassland. As a result, the range of many grassland species has expanded, whereas that of other species has contracted. There are far more red kangaroos in Australia now than there were when Europeans arrived in 1788. So once punctuated equilibrium is stripped of its radical misreadings, how does it collide with gene-selectionist neo-Darwinian orthodoxy?

Gould and Eldredge are right in thinking that the processes we can observe in local populations do not tell the whole story. That picture needs to be supplemented. The problem is not stasis but speciation. How can events in a local population generate a new species? This question is at the heart of Eldredge's recent book *Reinventing Darwin,* which revisits the debate with Dawkins. The short answer is that they usually do not. Local populations change, as the example of Australian rabbits and myxomatosis shows. But changes in a local population are usually too fragile to make a new species. Adaptation to local circumstances as, say, an impala population adapts to a particularly dry region, usually depends on gene complexes rather than single genes. Such complexes in local populations are vulnerable to being swamped, either by migration or by one population fusing with another. Local populations are short-lived and their boundaries are permeable, and so the clock of local evolutionary change is always in danger of being set back to zero. The facts which make stasis easy to explain make speciation hard to explain.

Nonetheless, speciation is obviously possible. New species do come into existence. There are different ideas on how this puzzle is to be solved. But any solution will take us beyond events in local populations observable on human timescales. For example, Elizabeth Vrba (another of Gould's co-authors) argues that occasional climate changes are responsible for a 'turn-over pulse'. These changes rob some species of their entire habitat, and those species go extinct. But other species will be fragmented. Some of their populations may change their character. Instead of being semi-isolated, or briefly isolated, they will be fully isolated, and for long periods. Most such isolated fragments will sooner or later go extinct. But a few will become new species. For changes within them will accumulate rather than being washed away by fusion within the larger population.

Vrba's particular model might have only partial validity. But it is likely that whatever explains the occasional transformation of a population into a species will rely on large-scale but rare climatic, geographic or geological events; events which isolate populations until local change is entrenched. This is an exception to extrapolationism. We cannot understand speciation just by studying evolutionary change in local populations. But it is not a radical break. Dawkins could, should and probably would accept it. After all, Ernst Mayr, one of the architects of contemporary Darwinism, has long defended a view of speciation along these lines. Gould somewhat overstates the adherence of orthodoxy to strict extrapolationism. Punctuated equilibrium is more important than the rather ungenerous treatment that Dawkins gave it in *The Blind Watchmaker*. For he interpreted it as an idea about the rate of change in local populations. I see it as a thesis about how, and under what circumstances, local changes become speciation events. If these circumstances are unusual, and if, as Vrba and Eldredge argue, there is a common thread to them, then the theory of speciation is an ingredient we need to add to the gene selectionist's tool kit. Ernst Mayr taught evolutionary biologists that speciation occurs only when a fragment of an ancestral species becomes geographically isolated from the rest of the population and diverges from that ancestor. Vrba and Eldredge add to Mayr's original insight a theory of when such fragmentation is most likely to take place, and when the fragments are likely to survive rather than go extinct. Punctuated equilibrium is, then, an important idea.

9 Mass Extinction

Extinction is normal. Species go extinct as a result of local ecological interactions. The Stephens Island wren lived only on (surprise!) Stephens Island (in New Zealand), and is now extinct as a result of wren/cat interactions. Other species disappear as a result of competitive displacement. Still others are just unlucky; they are unfortunate enough to be on a volcano that erupts or in a lake that dries out. These examples pose no problem for the view that the evolutionary history of species and species lineages is just an aggregate of local ecological processes of the kind we can and have observed. For we do observe hunting cats, erupting volcanoes and drying lakes. However, Gould argues that many extinct lineages did not die the death of a thousand microevolutionary cuts. They have gone out not with a myriad of whimpers but as part of a big bang. Major lineages in the tree of life typically go extinct in periods of mass extinction; periods which change the rules of the evolutionary game.

Earth history is divided into eras, periods and epochs

largely by the contrast in their species composition. Thus the divide between the Permian and the Triassic, about 260 million years ago, is one of the most profound divides of earth history. It marks the close of the Palaeozoic era and the beginning of the Mesozoic. Douglas Erwin's *The Great Paleozoic Crisis* begins with a 'before and after' snapshot of typical marine invertebrate communities across this boundary. The shift is dramatic. The Permian community is dominated by filter-feeding animals attached firmly to a base. Most animals are immobile. There are exceptions: mobile animals like fish, cephalopods (squid, octopus and their relatives), snails and bivalves (clams, oysters, mussels and the like) are part of the community. But they are present only in small numbers. In contrast, Mesozoic communities are dominated by animals which move under their own steam. Most of the fixed-in-place animals of the Permian, together with the coral species which built the reefs on which the former established themselves, have gone. The biological world had changed massively.

These divisions in earth's history seem to imply the existence of some great killing episode. For if there were not a particular event, *a specific killer*, why would deaths in, say, the snail lineage (the gastropods) be correlated with those of the starfish and sea urchins (the echinoderms) or with terrestrial reptiles? The very organisation of geological history seems to presuppose a

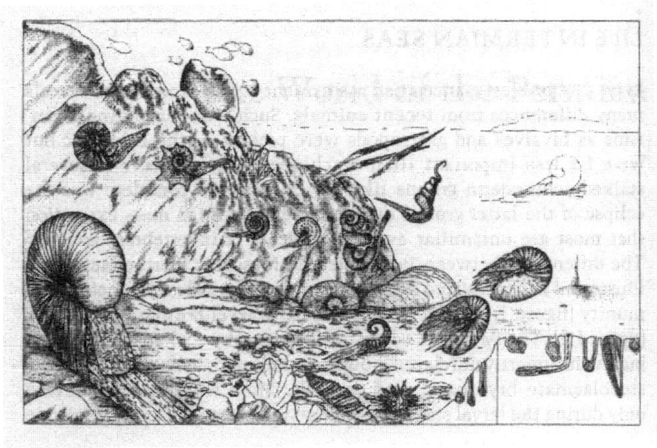

Figure 3: The advent of the Mesozoic era saw an explosion in the number and diversity of mobile taxa in marine communities. The first picture shows a Palaeozoic reefal community, the second a late-Mesozoic bottom community. (Source: Drawn by Mary Parrish, in D.H. Erwin, *The Great Paleozoic Crisis*, New York: Columbia Universitics Press, 1993.)

real difference between successive biotas, a difference with a relatively unified cause. If so, massive extinctions must be of profound importance, reshaping the tree of life. In these episodes, major *clades* – founding species and all their descendants – disappear. The trilobites disappeared in the end-Permian extinction, and the ammonites are never found beyond the Cretaceous/ Tertiary boundary. But even those clades that survive are often profoundly altered. Erwin points out that though the echinoderm and snail lineages survived the Palaeozoic crisis, only a few species in each survived. Their diversity was profoundly reduced, and that reduction in diversity marked the rest of the history of those groups. A 'Modern fauna' can be snail rich, but still unlike the Palaeozoic snail fauna. In Erwin's view, the Palaeozoic crisis reshaped the whole history of life.

As always, matters are not so simple. The trilobites disappeared for ever at the end of the Permian. But the diversity of the trilobite lineage had already shrunk profoundly before that disaster. Some defend similar views of the dinosaurs, arguing that their range and biological diversity had already shrunk before the end of the Mesozoic. If this line of thought is right, and mass extinctions merely speed up a process already underway, then they would not make much difference in the very long run. If mass extinction's effects are selective, and the less well-adapted species are those most likely to go extinct, major crises – periods when many species die – might simply accentuate evolutionary trends already in process. David Raup calls this the 'fair game' model of mass extinction.

In evaluating the 'fair game' model, it is important to discover the nature and duration of mass extinction episodes. If they are genuinely sudden – sudden in an ecological time frame, as catastrophic climate changes caused by an impact would be – then selective accounts of extinction become very implausible. For the survival of the species would depend on the suite of biological characters it happened to have at the moment of change; whether these are fortunate or unfortunate. But if mass extinction episodes take place over millions of years, then extinction can be sensitive to the evolutionary response of the lineage. The more structured, gradual, and smeared out over time a mass extinction is, the more likely it is that the distinction between global event-driven mass extinction and background extinction is a difference in degree, not kind. If mass extinction episodes are fast and discontinuous with the events that surround them, then mass extinctions will have distinctive evolutionary effects. They will change, perhaps profoundly, the history of life.

This issue has been debated most vigorously in connection with the death of the dinosaurs. No one now seriously doubts that there was a meteor impact at the

Cretaceous/Tertiary boundary. But there is still a lot of debate on its significance. After all, if that is all that happened, why did crocodiles, turtles and even frogs go through relatively unscathed? One line of argument suggests that the ammonites, together with dinosaurs (other than birds), pterosaurs, pliosaurs and other marine reptiles had all shrunk significantly in diversity and range before the impact. Perhaps some of these groups were already extinct, and so talk of the 'Cretaceous/Tertiary extinction' compacts the process in time. It treats events that had been happening over millions of years as if they happened in a geological instant. Others argue that the great clades that failed to make it to the Tertiary were in good shape before unforeseeable catastrophe overtook their world.

In the case of dinosaurs, perhaps the meteorite only administered the final blow to a group on the way out. But I do not think that this can be in general true of mass extinction. The changes they impose are too vast. That is particularly true of the catastrophe that struck life at the end of the Permian. Probably more than 90% of animal species then alive went extinct. Extinction on this scale must have caused fundamental reorganisations of life. If so, we cannot understand the overall history of life by projecting, onto the largest scale, processes we see operating in local populations. Mass extinctions are not just local bad news scaled up.

Moreover, Gould, drawing on the work of David Raup, argues that there is a distinctive evolutionary regime in operation in periods of mass extinction. These are not casinos ruled by chance alone. There are principles which would enable us to pick winners and losers. The game has rules. But they are different rules from those of normal times. The magnitude of the upheaval at the Permian/Triassic boundary, and the pace of the upheaval at the Cretaceous/Tertiary (if the impact was important), make it unlikely that the game was fair. Adaptation, we recall, is adaptation to a specific environment. Scramble the environment – drop a polar bear in the desert – and even a species superbly adapted to its previous environment will be in deep trouble. So, as Raup puts it, extinction was probably 'wanton'. Species survival is not random, but the properties on which survival depends are not adaptations to the danger mass extinction threatens. If a meteor impact caused a nuclear winter, then the ability to lie dormant would have improved your chances. But dormancy is not an adaptation to the danger of meteor impacts.

The ability to lie dormant is a characteristic of individual organisms. But many important characteristics relevant to survival or extinction would have been properties of species themselves. Species with broad geographic ranges, species with broad habitat tolerances, species whose lifecycle does not tie them too closely to a particular type of community all would have had a better chance of making it. At least, it is plausible to suggest this, though testing these suggestions empirically turns out to be very difficult. In any case, Gould argues that survival and extinction through periods of mass extinction involve some form of species selection. If so, mass extinctions are doubly important. They restructure the history of life, and they do so in part through a sieve, a filter, not visible in local populations in local communities. For filters in local communities are sensitive to the properties of organisms, not species.

In short, Gould's case for the importance of mass extinction depends minimally on the view that there is a qualitative difference between mass extinction and background extinction, and that major groups have disappeared that would otherwise have survived. While hard to prove, that claim is very plausible. It depends further on the idea that species-level properties in part determine survival. Mass extinction regimes are species selection regimes. Once more, that is a plausible conjecture, but it awaits clear confirmation.

10 Life in the Cambrian

Conventional wisdom emphasises the gradualness of evolutionary change. New organs – circulatory systems, nerve nets, limbs and tentacles, perceptual organs – are put together bit by bit over countless generations. So are new ways of organising tissues and organs into functional animals. In this respect, Dawkins is a true son of orthodoxy. He cannot remind us too often that the power of selection to build our exquisite and intricate biota depends on its slow and incremental operation. Every living creature is a triumph over chance. No random process, no hurricane blowing through a junkyard, could ever assemble anything as wildly improbable as a flea or a weevil. Each organic design is a victory over the improbable, and each is won by insensible degree. Mount Improbable is ascended by the smoothest and gentlest of tracks.

This standard story seems to run slap-bang into a nasty fact. About 530 million years ago, the fossil record seems to show that most of the major animal groups appeared simultaneously. In the 'Cambrian explosion', we find segmented worms, velvet worms, starfish and their allies, molluscs (snails, squid and their relatives), sponges, bivalves and other shelled animals appearing all at once, with their basic organisation, organ systems, and sensory mechanisms already operational. We do not find crude prototypes of, say, starfish or trilobites. Moreover, we do not find the common ancestors of these groups. Multi-celled animals are probably a monophyletic clade: a single ancestral species that gave rise to all, and only, the animals. Modern groups arose from this common ancestor. So there must have been animals that were, for example, the ancestors of the arthropods, segmented worms and velvet worms. Since those animals share the basic pattern of segmentation, they all probably descended from a segmented ancestor. But no likely candidate has been found in the fossil record.

This abrupt and explosive evolutionary radiation of the Cambrian seems to be unique. Plants seem to have arisen more gradually. Flowering plants evolved well after the gymnosperms, and gymnosperms were proceeded by earlier plant lineages, some now wholly extinct. Nor was there a similar radiation when animals invaded the land. Dry land – and even fairly moist land – provided an empty ecological space for the first animals which could adapt to living out of water. But the

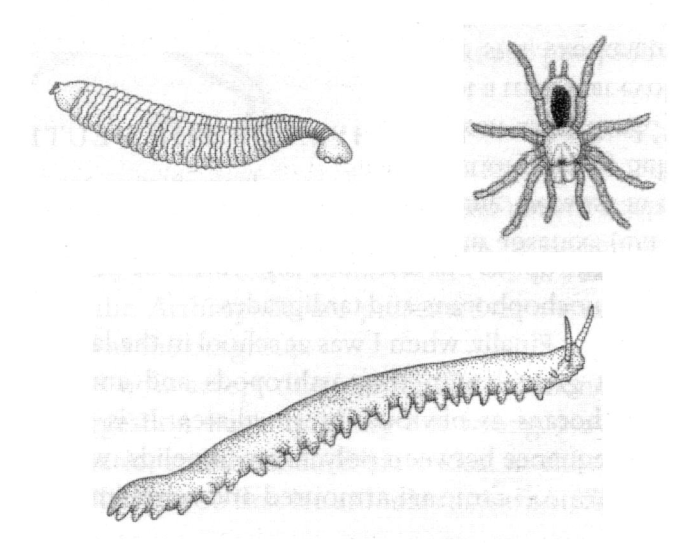

Figure 4: The 'Cambrian explosion' saw a host of new creatures appear at the same time, perhaps challenging Dawkins' view of a gradual evolutionary process. This segmented worm (leech), arthropod (spider) and velvet worm all arose *c*. 530 million years ago. (Source: C. Tudge, *The Variety of Life,* Oxford University Press, 2000.)

colonisation of the land saw no proliferation of wholly new ways of making an animal. Arthropods remain recognisable arthropods; and snails, snails. Vertebrates retain the basic vertebrate organisation. Worms, too, stay worm-like, even as representatives of all these groups acquired the special adaptations needed for life in the dry.

So perhaps evolution worked under different rules

back then. As we shall see, Gould is inclined to think so. One alternative is to argue that the 'explosive Cambrian radiation' of animal life is an illusion generated by the failure of earlier pre-Cambrian fossils to survive to our times. Darwin and many of his successors supposed that the emergence of the animals only *seemed* explosive. They thought the first appearance of multi-celled animals in the fossil record was preceded by a long history of hidden evolution. This remains a live option. Shortly after World War II, fossils of pre-Cambrian animals were first discovered in Ediacara, South Australia. Subsequently it has been shown that this Ediacaran fauna was present worldwide. So there was animal life before the Cambrian. However, the relationship between the Ediacaran fauna and that of the Cambrian remains very unclear. In one view – a view Gould is inclined to support – there is no relationship. The Ediacaran fauna were a failed experiment in the history of life; a branch of the tree of life that was wholly extinct before or at the Cambrian. The Ediacaran animals were not ancestors of the Cambrian animals and hence their existence does not extend the time-frame of animal evolution into the Pre-Cambrian.

Ediacaran fossils are not the only reason for suspecting the existence of a hidden history that pre-dates the Cambrian explosion. The last few decades have seen the development of molecular methods of estimating the time at which two lineages diverged. When, say, the velvet worm lineage diverged from that of the arthropods, each lineage inherited its genetic material from that last common ancestor. Once the lineages began to evolve independently, differences started to appear in their genetic material. If we can measure the extent of that difference, and calibrate the rate at which DNA sequences diverge from one another, we can estimate the time the last common ancestor lived. The idea is to compare a stretch of DNA in the velvet worm lineage with an equivalent stretch in an arthropod, and measure the extent to which they have diverged from one another. If the rate of divergence can then be calibrated from lineages with rich, detailed fossil records, we can then estimate the time when the last common ancestor of the velvet worms and arthropods lived. This technique has many potential uncertainties. It depends on careful calibration of the rate of change of genes. It depends on careful choice of the genes used. For example, suppose we were to choose the genes that code for the neural networks of the two animals. If those genes had been subject to strong selection in, say, the arthropods, then that selection would have produced genetic change. Arthropod neurone-making genes would have become quite different from velvet worm neurone-making genes. If selection on arthropods had been more intense than on the lineage we used to calibrate the 'genetic clock', we would overestimate the depth in time of the divergence of the two lineages.

The strength of selection, and hence the rate of change it causes, can obviously vary over time and between lineages. So these uncertainties can be reduced by choosing genes which are not being changed by selection. Many genes are silent, coding for no protein at all. A change in a silent gene has no effect on an organism's phenotype. Such mutations are 'neutral', and are not visible to selection. The same is true of changes to a gene which make no difference to the protein it builds. Such changes are possible because the gene/protein code is redundant. The protein-building machinery reads DNA sequences in sets of three bases, and quite often a change in the third position makes no difference to the protein that is made. The general presumption with genetic clocks is that the rate of neutral changes will not vary much between lineages or over time. So we can increase our confidence in molecular clocks by choosing silent genes or genes which code for highly-conserved features of organisms; for example, for very basic metabolic functions common to all animals. Neither type of gene will be changing as a result of selection. A second way is to use a number of DNA sequences not just one. If two, three or more different clocks all give similar dates of divergence, we can be much more confident that they are approximately right.

Information from molecular clocks increasingly suggests that the main animal lineages of the Cambrian had last common ancestors that pre-date the Cambrian by hundreds of millions of years. The last common ancestor of, say, the trilobite lineage and the flatworm lineage may have lived over 800 million years ago. Despite the uncertainties of molecular clocks, it is increasingly clear that the animal phyla first recorded in the Cambrian did indeed have a hidden evolutionary history. The last common ancestor of the animal phyla lived a very long time before the Cambrian. Gould accepts this, but rightly points out that it remains possible that the Cambrian explosion really was explosive. The divergence of two lineages is one thing; the acquisition of their distinctive organisation and equipment is another.

In Figures 5 and 6, the horizontal bars indicate evolutionary changes in the lineage. So Figure 5 depicts a history in which those distinctive morphologies are built step by step from the time of initial divergence. Figure 6, on the other hand, depicts an evolutionary history in which the velvet worm and arthropod lineages diverged in the deep past, but developed their distinctive morphologies in a rapid evolutionary burst long after their initial divergence. Gould points out that molecular clock data cannot decide between these two possibilities. Moreover, the fossil record supports the Figure 6

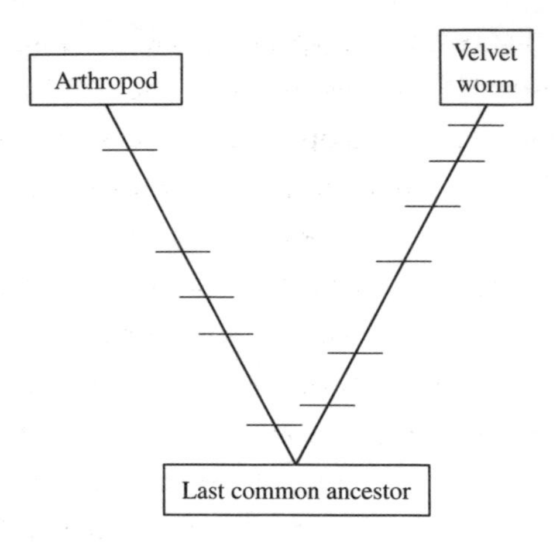

Figure 5: A history in which the distinctive morphologies of velvet worm and arthropod lineages are built step by step from the time of initial divergence.

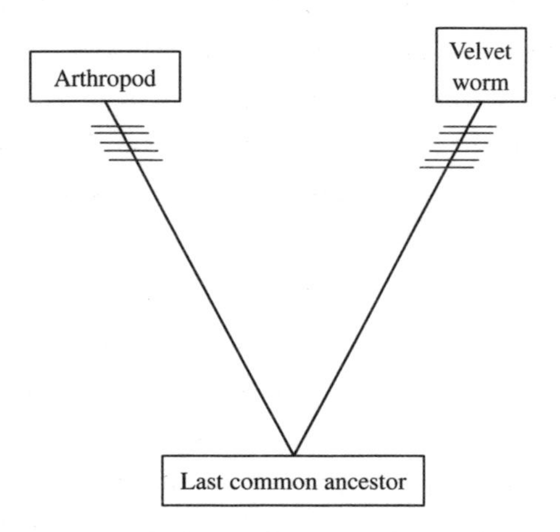

Figure 6: A history in which the distinctive morphologies of velvet worm and arthropod lineages develop in a quick burst, long after the initial divergence.

picture; for example, it explains the lack of Pre-Cambrian proto-arthropod fossils. In short, the 'hidden history' hypothesis remains an open option, but so does Gould's guess that the Cambrian explosion was genuinely explosive rather than an illusion generated by incomplete preservation. Evolution at the base of the Cambrian may really have been especially fast. Perhaps competition was weak in an empty world. Perhaps the developmental programs of early animals were more open – less constrained – and so more variation was generated.

Gould's ideas about the rapidity of the early Cambrian radiation form the least controversial of his claims about this era of life. For he thinks recent evidence about the Cambrian fauna overthrows our standard conception of life's history, a picture that sees life as becoming more diverse and better adapted over time. As he sees it, a remarkable discovery made in the early years of the twentieth century overthrows this conception. The discovery was that of the Burgess Shale fauna; a discovery remarkable because the Burgess shale preserved not just bits of shell and bone, but soft structures as well. So it yielded a record of creatures without hard parts, creatures whose existence would otherwise never have been suspected. And it reveals much more than we would usually know about those creatures with hard bits.

The distinguished American palaeontologist, Charles Walcott, discovered the Burgess Shale fossils, but interpreted these creatures as simpler versions of well-known types of animal. During the 1980s, this interpretation was radically revised. The revision suggested that many of the Burgess animals were radically unlike anything living. Some were recognisable arthropods. But they were not members of any of the four great arthropod groups: spiders and their allies; trilobites; crabs, lobsters and the like; or insects and insect-like forms. So, the idea ran, the Cambrian saw not just the invention of the arthropods – segmented, jointed-limbed, exoskeleton-covered animals – but of far more kinds of arthropods than have ever been seen since.

As we saw in Chapter 7, in developing this idea, Gould distinguished between the *diversity* of life, and its *disparity*. The diversity of life is the number of species extant at that time, and no one doubts that the diversity of life has increased since the Cambrian. Disparity is not counted by species numbers. Disparity measures the morphological and physiological differentiation between species. The discovery in recent years that New Zealand has not one but two species of *tuatara* was the discovery of extra diversity in New Zealand. (The tuatara is a lizard-like reptile; the sole living survivor of the sister group to snakes and lizards.) But it was not the discovery of extra disparity. For the two species are so alike that the existence of separate species was only suspected when molecular techniques showed different populations on different islands were genetically quite dissimilar, even though structurally they are almost identical. However, the nineteenth-century discovery in New Zealand of the tuatara, like that of the echidna and platypus in Australia, was the discovery of significant additional vertebrate disparity. For the *monotremes* are unlike other mammals not just in laying eggs. They are structurally unlike other mammals, too, in having a single opening that serves for both reproduction and excretion.

Armed with the distinction between disparity and diversity, we can now explore Gould's more radical claims about the history of animal life. As Gould sees it, though diversity has increased, disparity has radically shrunk since the Cambrian. Animal life was at its most disparate at the peak of that explosion. The arthropods are by far the largest clade of animals, and there were more fundamentally different kinds of arthropod alive in the Cambrian than the world has seen since. The same is true at an even grander scale. The major subdivisions of animal life are phyla. Each phylum is a distinctive way of building an animal. The molluscs, for example, jointly form a single phylum. Amongst the Burgess Shale fauna, Gould says, we find many animals that are members of no surviving phylum. And they are as different from the modern phyla – from the velvet worms, arthropods, molluscs, vertebrates, flatworms, starfish, bivalves and the others – as each of those phyla are from one another. To put the point succinctly: there were many phyla alive then that are no longer around. Some phyla consist of tiny soft-bodied animals and so have next to no fossil record. But, with a single exception, all the living phyla that have reasonable fossil records are found in the Cambrian. So the Cambrian phylum count was larger, maybe much larger, than the contemporary count. No new phyla have appeared, and many have gone. That count, in turn, is a reasonable measure of disparity. So Cambrian disparity was considerably greater than current disparity. The history of animal life is not a history of gradually increasing differentiation. It is a

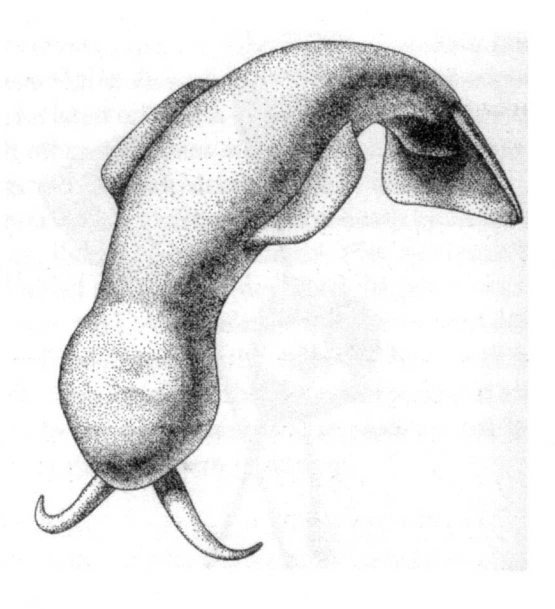

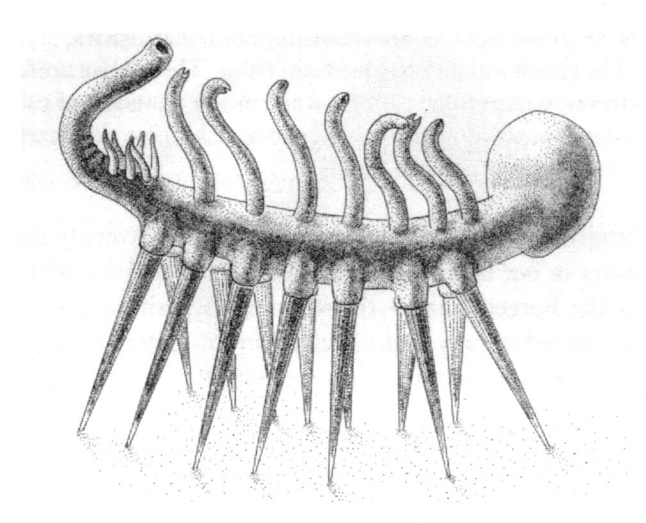

Figure 7: The flattened swimming animal *Amiskwia* (opposite), with a pair of tentacles on the head, and side and tail fins behind. *Hallucigenia* (above), supported by its seven pairs of struts, stands on the sea floor. (Source: Drawn by Marianne Collins, in S.J. Gould, *Wonderful Life,* London: Hutchinson Radius, 1990.)

history of exuberant initial proliferation followed by much loss; perhaps sudden loss.

If all of this is right, it raises some very fundamental questions about life's history.

Why was the Cambrian so rich in disparity, and why was the disparity generated so rapidly, and then lost? Gould rather doubts that selection has much to do either with the early burst of disparity or with the roster of loss and survival. Perhaps most pressingly of all: why has there been so little disparity generated since the Cambrian? If no or few phyla have evolved since the Cambrian, and if counts of phyla measure the disparity of life reasonably well, the history of animal life since the Cambrian has been strikingly conservative. No new plans have been invented; no old ones have been massively modified. If Gould is right about this basic pattern of history, he is surely right in thinking we are faced by a mystery. If early animal life was highly disparate, and if little new disparity has subsequently evolved, we need to know why. For evolutionary change in general has not ground to a halt in the last 500 million years. That period has seen the evolution of all the adaptive apparatus needed for life on, under and above the ground. Many adaptive complexes, not least human intelligence, have been invented in that period. So why have new phyla – major new ways of organising animal bodies – not been invented too?

Dawkins and, even more, his former student Mark Ridley think the basic claim about the pattern of history is not right. They challenge Gould's views in two ways. Ridley, in particular, is sceptical of the distinction on which the whole picture rests. He is a committed cladist. Cladists have a very distinctive conception of the aim of biological classification. They think that a biological classification is an evolutionary genealogy. The 'cladogram' in Figure 8 makes no claim at all about the morphological, physiological or behavioural similarities

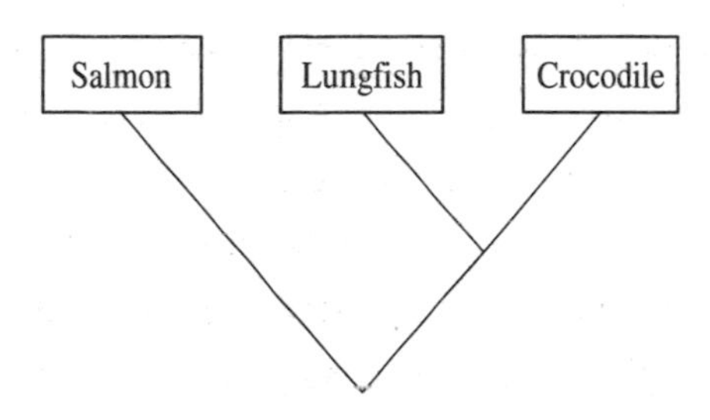

Figure 8: Lungfish are more closely related to crocodiles than they are to salmon, yet systematicists once classified lungfish and salmon together on the basis of morphological similarities.

of salmon, lungfish and crocodiles. Rather, it claims that lungfish and crocodiles share a more recent common ancestor than either share with the salmon. The purpose of biological systematics is to discover and represent genealogical relationships between species. And that is its only purpose. Cladists think that the only groups we should recognise and name – genera, families, orders, classes and phyla – are *monophyletic*. A monophyletic group consists of *all* and *only* the descendants of a single founding species. Mammals are monophyletic: a single ancestral species gave rise to all and only the mammals. The group of animals we call reptiles, by contrast, is not monophyletic. There is no species that had all and only the reptiles as its descendants. Birds and mammals also descended from the common ancestor of all the reptiles. So cladists do not think of the reptiles as a real group.

The cladists' reasons for focusing on genealogical relations strike at the very idea of disparity. They do not think we can objectively measure similarity. Similarity and dissimilarity are not objective features of the living world. Our judgements of similarity and difference reflect the biases of human perceptions and interests, not the objective features of the world. We are visual organisms; sight is our primary sense. We are therefore very struck by differences in visual appearance. One of Gould's weird wonders of the Burgess Shale, *Opabinia,* has a cluster of five eyes. It certainly looks genuinely odd. But imagine we were equipped with noses like a bloodhound, and that scent was as rich a source of information for us as vision. Perhaps we would then think of the moths, say, as varying wildly from one another, in virtue of differences in the pheromones they use to attract mates. Organisms differ from one another in their morphology and physiology in endless ways. Some of those differences are more salient to us, more striking and surprising, than others. But according to the cladists, that is a fact about us, and the ways we interact with the world. *It is not a fact about the history of life*. A sentient electric eel, given the same data, would reconstruct the genealogy of life – who is related to whom – in the same way we do. For the pattern of descent is an objective fact of history, though one difficult to discover. But would they make the same judgements of disparity? Cladists doubt it.

Like many evolutionary biologists, Gould uses spatial metaphors to explore and explain his ideas, and he recruits a spatial metaphor to explain disparity. 'Morphospace' is the space of animal designs. In Chapter 6 we saw a cut-down region of morphospace, a three-dimensional space which represents all possible shells. This three-dimensional space enabled us to represent shell disparity. It enabled us to represent all the possible ways shells might be shaped, and it turned out that real shells occupy only a small region of that space. To represent the disparity of real organisms, even those that live in shells, we need a morphospace with many more dimensions. Three dimensions do not, for example, capture the contrast between oysters and brachiopods. Both are shelled animals, but brachiopods have a very distinctive feeding apparatus (and are notoriously inedible). Three-dimensional shell morphospace tells us nothing about the organisation of the animal in the shell. One way of putting the cladists' point is to ask: how do we determine the dimensions of morphospace? How do we measure distance in any given dimension? We can measure an infinite number of features of any given animal. We can measure the number of leg hairs on a fruit fly. Is this a dimension of morphospace? What about the ratio of the facets of its compound eye to their leg hair number? If that sounds absurdly recondite, it is not. Insect taxonomists routinely tell one species from another by measuring the curves, crinkles, hooks and barbs on their sexual organs. For disparity to be an objective feature of the tree of life, there must be a principled way of answering questions such as these. There must be some principle that shows, for example, that variation in leg number amongst arthropods is a genuine aspect of disparity, whereas variation in nostril hair number in primates is not. Cladists doubt that any such principle is to be found. Gould accepts that this challenge is difficult to meet, but believes palaeobiology can and must develop ways of meeting it.

Dawkins develops a different challenge. He argues that even were we to accept that Gould's basic distinction between diversity and disparity were sound, Gould over-counts Cambrian disparity. To understand Dawkins' point, we need to make a short venture into arthropod taxonomy. The arthropod trunk of the tree of life divided into four great branches – four classes – that, except for the extinct trilobites, have been evolving independently of one another for 500 plus million years. These classes are the trilobites, the insect-like arthropods, the crustaceans, and the spider-like arthropods. Systematicists identify these animals on the basis of the pattern of bodily segments, and the number and pattern of limbs and feelers on those segments. Living arthropods exemplify three basic patterns. Crustaceans, for example, have a basic division into the head and the trunk. The trunk varies a good deal, but the head is always divided into five segments, each with a pair of branched limbs. Of these five pairs, two are antennae, two are maxillae, and one is mandibles. Systematicists are interested in these features because they are good markers of the genealogical relations between arthropods. The evolutionary genealogist needs traits that show some, but not too much, evolutionary plasticity relative to the lifespan of the group of interest. The possession of an exoskeleton or the habit of laying eggs are important traits. But they are uninformative: they are universal amongst the arthropods. They are *too conservative*. Other traits, like the number of segments which form the body of the animal, are *too variable* for keeping track of the basic pattern of arthropod relationships.

It turns out that the basic pattern of body segmentation, the pattern of limb development (in particular, whether limbs develop as single or as branched structures) and the number of segments forming the head are features of arthropod history that have been evolutionarily conserved over periods of 500 million years. They are conservative enough, but not too conservative. Once an arthropod lineage has evolved, say, with the crustacean pattern of limbs on its head, animals later in that lineage do not lose that pattern. And animals outside that lineage do not evolve the trait independently. So it acts as a membership badge for that branch of the arthropod family. In contrast, over long periods the possession of eyes is not a good marker of evolutionary relationships. For eyes are lost as well as gained, and they have often evolved independently.

This excursion does have a point! Gould takes traits – segmentation and appendage patterns – that are genuinely important as markers of genealogical connection amongst the arthropods, and treats these as measures of arthropod disparity. He identifies extraordinary levels of arthropod disparity in the Burgess Shale fauna on the grounds that this fauna exemplifies segmentation and appendage patterns not found in the four great branches of arthropod life today. So what? As we have just seen, these traits are not chosen for their intrinsic importance. They are chosen because they have become conservative over periods of 500 million years. There may be no special significance in crustaceans having five pairs of appendages on their heads. That may be just a minor but frozen historical accident. That pattern still indicates relatedness. The point is familiar from human genealogy. An unusual surname is of no intrinsic significance, but it can still indicate a family connection. So even if, contra Ridley's suspicions, disparity is a genuine property of the tree of life, there is no reason to suppose that it is measured by traits which are suitable for keeping track of relatedness over long periods of time.

The distinction between disparity and diversity is very

plausible. Some of the Burgess animals really do look strange and wonderful. Even so, it is fair to say the Ridley/Dawkins challenge is yet to be met. We lack a good account of the nature of disparity, and we lack objective measures of it. Without that, the existence of Gould's puzzling pattern remains conjectural.

11 The Evolutionary Escalator

Three and a half billion years ago, the most complex and sophisticated life forms were cynobacteria. Cynobacteria are not just single-celled organisms. They are single-celled organisms lacking a nucleus, mitochondria, chloroplasts and a raft of other internal structures. Many eukaryotic single-celled organisms now exist. These are organisms with parts very nearly as complex as a bacterium (probably because they once were bacteria). Still more obviously, multi-celled plants and animals have evolved. These are not just huge assemblages of cells. They are differentiated assemblies. Animals consist of an array of cell types which are organised into tissues, organs, organ systems, and so on. This is an astounding evolutionary achievement. As an animal or plant grows from a single fertilised cell, it does not just get larger through cell division. As the cells in an animal divide, at some stage they have to begin turning into neural cells, muscle fibres, blood cells, sex cells, and the tissues of an array of specialised organs. Eyes, for example, need cells with photoreceptors. The specialised cells have to be assembled into larger structures – tissues and organs – and connected appropriately with others.

All this has to take place while the embryo remains functional. Functional enough, at any rate, to stay alive and, in many species, to fend for itself. Millions of different developmental packages have evolved in the last billion years or so, and the result is an extraordinary array of multi-celled organisms. These are now capable of surviving and reproducing in an array of habitats from the highest mountains to the very deepest parts of the oceans. These habitats were invaded in stages by both plants and animals. Even after animals established themselves on land, it took time for desiccation-resistant eggs to evolve, to free reproduction from dependence on water. Plant evolution, too, showed a similarly incremental infiltration of terrestrial habitats.

In the light of all this, surely it is obvious that the history of life on earth shows a progressive increase both in complexity and adaptiveness. Gould does not quite reject this view outright. But he thinks it is a very misleading way to think about the history of life. His reasoning connects the direction of life's history with a major theme of his work: the importance of non-selective explanations of broad patterns in the history of life. Gould offers a reinterpretation of evolutionary trends. This reinterpretation includes the largest-scale trend of all: the tendency of life to become increasingly complex over time.

Let's begin with horses. As early as the nineteenth century, when Thomas Huxley championed Darwin's ideas, the history of horses was one of the paradigms of evolutionary change. In response to the opportunity provided by the evolution of grass, and the establishment of grasslands, horses became prairie animals rather than forest animals. So at least the standard story has it. But Gould thinks that this trend in horse evolution is a mirage. What has really happened in the horse lineage is a loss of diversity. There has not been a directional trend in horse evolution. Rather, there has been massive extinction in that lineage and the paltry surviving remnant happen to be grazers. The appearance of a trend is generated by a reduction in the heterogeneity of that lineage.

In discussing complexity, Gould tells a similar story at the very largest scale. What we think of as a progressive increase in complexity is a change in the range from the least to the most complex organism. It is a change in the spread of complexity. Life starts off as simple as life can be. Physics and chemistry impose constraints that define the least complex possible form of life. Bacteria are probably close to that limit, so life starts at the minimum level of complexity. Since even now nearly everything that is alive is a bacterium, for the most part life has stayed that way. But evolution will from time to time build a lineage that becomes more complex over time. No global evolutionary mechanisms prevent more complex organisms evolving from less complex ones. But none make it more likely. The complexity of the most complex life form alive tends to drift up over the generations, just because the point of origin of life is close to the physical lower bound. Relative to bacteria, there are never many of these complex creatures. But the difference between the simplest and the most complex organisms alive tends to become larger over time. If life originates close to the point of minimum complexity, wholly undirected mechanisms will increase this range. Mechanisms that are *blind to complexity* suffice for an *upward drift* in average complexity. But bacteria continue to dominate the living world. So it is misleading, at best, to think that evolution is characterised by a trend towards increased complexity.

Gould's picture is illustrated by the two snapshots of life in Figures 9 and 10. The first is of life soon after its origin. There is little variation in complexity. Everything alive is close to the lower bound – the left wall – of complexity. The second is a snapshot of life a few billion years or so later. The mode has not changed. Most living things are still close to the left wall. But the curve has

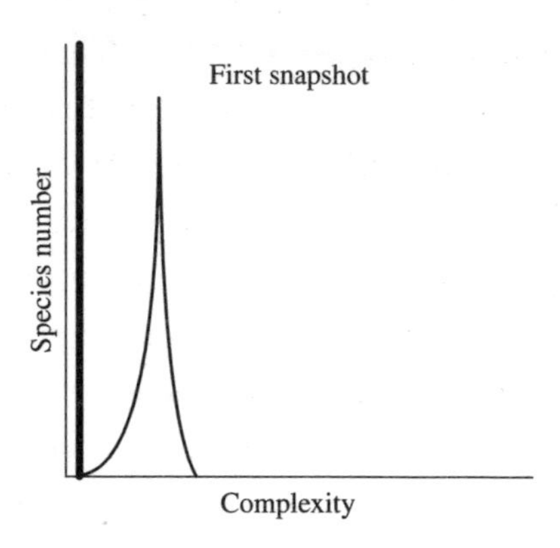

Figure 9: A snapshot of life soon after its origin. Here there is little variation in the complexity of organisms.

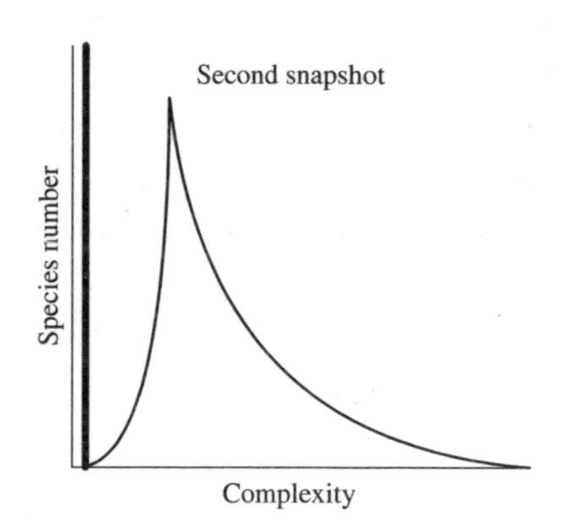

Figure 10: A second snapshot of life a few billion years later, Here the variation in the complexity of organisms has increased.

spread to the right but not to the left. For there is a wall imposed by the laws of the physical sciences to the left, but none to the right.

How might we respond to the view that the trend to increased complexity is a mirage? In Dawkins' view, complexity is a red herring. Both he and Gould are most concerned with progress. Gould is interested in complexity only because he regards complexity as a measurable indicator of progress. Dawkins does not – he thinks our interest in complexity is anthropocentric. He thinks evolution is progressive because, over time, life is becoming *better adapted*. As time goes on and natural selection grinds away, living creatures become better designed. They become better adapted to their niches. Organisms at one point in time are typically better adapted than earlier ones and not as well adapted as later ones. Over relatively short time periods, and when we are considering members of a single evolving population, this is uncontroversial. If two organisms are members of the same population responding to selection, we can certainly compare their fitness. But if we define progress as increasing levels of adaptedness over millions of years, then we are required to compare creatures with very different bodies, living in different environments. To make that comparison, we would need to be able to define a measure of 'fit' between an organism and its niche. The idea that there is such a measure has great intuitive force. However, despite its plausibility it has turned out to be very difficult to work out how to compare the fitness of different organisms living in different environments.

That is not surprising. Natural selection generates adaptation to the *local conditions of life*. While those conditions remain, we can say what would increase adaptedness. As bittern camouflage patterns improve, bitterns become better adapted. But this example depends on the fact that the niche of the bittern remains the same. Both the features of the environment and the relevance of these features to the bittern population are constant over an evolutionary transformation from less well-camouflaged to very well-camouflaged bitterns. There is evolutionary progress in a single lineage in a fixed environment. But the conditions of life are not stable over the long term. The physical parameters of environments change. Moreover, organisms come to depend on different features of their environment. If a mammal comes to be able to synthesise vitamin C (as most can), its needs change. The aspects of an organism's environment that matter to it change. The sceptic argues that we can only define progress in the short run. We cannot really compare the level of adaptedness of genuinely different plants and animals.

Dawkins disagrees with this pessimistic assessment of our capacity to identify progress. He argues that arms-race evolution between competing lineages defines an arrow of progress, an arrow of improvement over the long term, though not the very long term. Arms races between lineages are cut short by mass extinction events, but while they are in progress each lineage is objectively improving. I am not convinced by this view. Arms races change the environment in which evolutionary change takes place. So I do not see how this idea solves the problem of comparing the level of fitness of different creatures in different environments.

Nonetheless, I think Gould also overstates his case. There is more to the history of life's complexity than a gradual increase in variance. In 1995, John Maynard Smith and Eors Szathmary published *The Major Transitions in Evolution*. Their view of life's history involves a series of major transitions and hence an inherent directionality. Some of these transitions involve the change from naked replicators to the first organisms. But they also think of the invention of eukaryotes; cellular differentiation and the invention of plants, animals and fungi; and the origin of social life as major transitions. Dawkins, too, has defended a similar series of major transitions, though not in as much detail. I think the difference between Gould, on the one hand, and Maynard Smith, Szathmary and Dawkins, on the other, is a difference in how they picture the spread of complexity. Gould thinks of complexity as having a lower bound but no upper bound, and these features of complexity are fixed by biochemistry, not by the course of evolutionary history. Over time, given that life originates near the lower bound of complexity, the spread increases as it trickles away from the limit imposed by the minimum possible complexity.

Maynard Smith and Szathmary do not regard the walls as fixed over time. Until the foundations of eukaryotic life were gradually assembled, there was also an *upper bound* on complexity. That bound was set by the intrinsic limits on the size and structural complexity of prokaryotes. For much of its history, for perhaps 2 billion years, bacterial evolution was confined between these two limits. Similarly, after the evolution of the eukaryote, there was a shift of the upper limit, but only a relatively small one. The invention of the organism required a complex series of evolutionary innovations. Until these came into existence, there was an upper bound to complexity set by the limits on a single eukaryotic cell. Maynard Smith and Szathmary argue that social existence, too, has evolutionary preconditions. Until these are met, a wall remains to the right. So, where Gould sees unchanging boundaries set by physics and chemistry, Maynard Smith, Szathmary and Dawkins see evolution as transforming these boundaries irreversibly. The eukaryotic cell, sexual reproduction and cellular differentiation all change the nature of evolutionary possibility. These possibilities have changed over time in a direction that increases the maximum attainable complexity.

In short, over time the rules of evolution change.

Evolvability has changed. For what can evolve depends on the developmental mechanisms that determine the variation available to selection. These have changed over time. For example, once cellular differentiation had been invented, new variation was made available to selection. These changes opened up new possibilities, especially the possibility of more complex life forms. In *Full House*, Gould insists that this age, and every age, is the age of bacteria. Bacteria are the world's most numerous organisms. They have the most disparate metabolic pathways. They may well add up to most of the world's biomass. All this is true and important. But it is not the whole truth. We live in an age in which many biological structures are now possible that were once not possible. That too is true, and important.

Part IV
The State of Play

12 A Candle in the Dark?

Dawkins and his allies really do have a different conception of evolution from that embraced by Eldredge, Lewontin and other collaborators of Gould. But those differences do not explain the undercurrent of hostility this debate has generated; hostility that surfaced dramatically in the *New York Review of Books* exchange. No doubt some of that hostility has a banal psychological explanation. People in general do not much enjoy being told they are wrong, especially in public. A somewhat prickly response is no great surprise. But I doubt whether this is the whole story. Dawkins and Gould mostly argue about issues internal to evolutionary theory. But they have very different attitudes to science itself.

Dawkins is an old-fashioned science worshipper (Here I line up with him, not Gould). Like all scientists, he accepts the fundamental Popperian point that scientific theory is always provisional, always open to revision in the light of new evidence and new ideas. And he accepts, of course, that in the short run human error and human prejudice can block our recognition of important evidence and good ideas. But Dawkins is wholly untouched by the postmodern climate of current intellectual life. For him, science is not just one knowledge system amongst many. It is not a socially-constructed reflection of the dominant ideology of our times. To the contrary: though occasionally fallible, the natural sciences are our one great engine for producing objective knowledge about the world. In many cases, we can be confident that received scientific opinion is right, or very nearly right. And that knowledge is liberating. In short, for Dawkins science is not just a light in the dark. It is by far our best, and perhaps our only, light.

Gould's take on the status of science is much more ambiguous. For one thing, he thinks some important questions are outside science's scope. He defends this idea in his recent work on the relationship between science and religion. On this issue, Dawkins' views are simple. He is an atheist. Theisms of all varieties are just bad ideas about how the world works, and science can prove that those ideas are bad. What is worse, as he sees it, these bad ideas have mostly had socially unfortunate consequences. Gould, by contrast, seems to think theism is irrelevant to religion. He interprets religion as a system of moral belief. Its essential feature is that it makes moral claims about how we ought to live. In

Gould's view, science is irrelevant to moral claims. Science and religion are concerned with independent domains.

Gould's views on religion are doubly strange. First, it seems extraordinary to overlook the innumerable claims about the history of the world and how it works that are made by the various religions. The claim that the world was created by a being with intentions and expectations seems to be a factual claim about the world, not a moral claim about what to do. Furthermore, those factual claims are often supposed to be the basis of a religion's moral injunctions. So they are not minor details of religious belief systems we can reasonably ignore.

Second, Gould seems to commit himself to a very strange conception of ethics. Does he think that there are genuine ethical truths? Is there genuine moral knowledge? Recent thinking on ethics has gone two ways on this question. Perhaps the main contemporary line of thought is to argue that moral claims are expressions of the speaker's attitude or intentions towards some act or individual. To call someone a scumbag, for instance, is not to describe a particular moral property of that person. Rather, it expresses the speaker's distaste of that person and their doings. The main alternative is to defend some version of 'naturalism'. From this viewpoint, moral claims *are* factual claims. They are based on facts, though typically very complex facts, about human welfare. Gould seems to commit himself to denying both these options. If 'expressivism' is right, there is no independent domain of moral knowledge to which religion contributes. Our moral utterances are not designed to describe objective features of the world, but are instead vehicles for expressing our attitudes and emotions.

Alternatively, if naturalism is right, science is central to morality. For it discovers the conditions under which we prosper. The appeal to religion has largely dropped out of the picture. For one thing, religions really do seem to make claims about the world, and ones that cannot be rationally sustained. For another, even if these claims were true, they do not seem to give us any *moral reason* for action. This point was made in classical Greek civilisation, and it can be condensed into a single question: 'Is torturing babies bad because God forbids it, or does God forbid it because it is bad?' Give the first answer, and you are committed to the bizarre view that it would be right, not just prudent, to torture should God command it. Give the second, and you concede the irrelevance of religion to moral truth.

So Gould thinks that there are important domains of human understanding in which science plays no role. Moreover, he is much more sceptical about the role of science, even in its 'proper' domain. Even so, he certainly rejects extreme versions of postmodern relativism. It is an objective fact of evolutionary history, and one that we know, that dinosaurs evolved by the Triassic, dominated terrestrial ecosystems during the Triassic, Jurassic and Cretaceous, and (with the exception of the birds) went extinct at the Cretaceous/Tertiary boundary, probably as a result of a large meteor striking the earth. There is no sense in which this is just a Western creation myth, a reflection of the dominant ideology of these times, or just an element of the current palaeobiological paradigm. It really happened that way, and we really know that it did. So *to some extent* Gould shares with Dawkins the view that science delivers objective knowledge about the world as it is.

Scientific belief is sensitive to objective evidence. It is more than a mere reflection of the culture and values of its times. But Gould argues that science is very deeply influenced by the cultural and social matrix in which it develops. Many of his *Natural History* columns illustrate both the influence of social context on science, and also its ultimate sensitivity to evidence. This sensitivity of science to its cultural location need not distort it. Sometimes the influence is beneficial, providing useful metaphors and models. Darwin's debt to nineteenth-century political economy is the most famous such example. In *Time's Arrow, Time's Cycle*, Gould locates the development of our conception of deep history in its cultural and intellectual context without any suggestion that that cultural context perverted the development of geology. But when the scientific issues are directly relevant to social and political concerns, all too often these sociocultural interests have led to bad science, pseudo-science, racist and sexist science. *The Mismeasure of Man* is Gould's most famous essay on these themes. In it he is concerned to show how a particular ideological context led to a warped and distorted appreciation of the evidence on human difference.

So, one sharp contrast between Dawkins and Gould is on the application of science in general, and evolutionary biology in particular, to our species. This is surely a source of much of the underlying tension in this debate. Perhaps a little surprisingly, Dawkins has not written systematically on this issue. And much of what he has written explores some of the differences between human evolution and that of most other organisms. For in human evolution, *memes* – ideas and skills – are important replicators. Ideas are copied generation by generation, just as genes are. Tunes, football allegiances, ethnic prejudices and skills are copied from one human to another. We humans are vehicles of the memes we carry, not just the genes we carry. This fact makes our evolutionary history importantly different from that of most creatures. For one thing, meme evolution is much faster than gene evolution. Even so, it is clear that Dawkins sees no problem, in practice or in principle, in applying evolutionary theories of social behaviour to humans.

The contrast with Gould is deep. Of course, Gould accepts that we are an evolved species. But everything Gould does not like in contemporary evolutionary thinking comes together in human sociobiology and its descendant, evolutionary psychology. The result has been a twenty-year campaign of savage polemic against evolutionary theories of human behaviour. Gould *hates* sociobiology. It is true that some evolutionary psychology does seem simple-minded. For example, Randy Thornhill's work on rape is unconvincing. He argues that sexually excluded men can in some circumstances improve their fitness by acts of rape, but he makes no attempt to take into account the fitness costs of sexual violence, and neglects obvious and serious problems for the idea that a tendency to rape is an adaptation. It is tempting to believe that *The Natural History of Rape* is a deliberate provocation.

Many contemporary evolutionary psychologists have taken on board the need for caution in testing adaptationist hypotheses. Certainly, Dennett repeatedly insists that we cannot assume that every characteristic is an adaptation. However, even those more cautious defenders of sociobiology, and its intellectual descendants, downplay the aspects of evolutionary process central to Gould's thought. They tend not to emphasise the importance of development and history in imposing constraints on adaptation, the problems in translating microevolutionary change into species-level change, the role of contingency and mass extinction in reshaping evolving lineages, or the importance of palaeobiology to evolutionary biology. Sociobiology, even at its most disciplined, reflects a different angle on evolution to that exemplified by Gould. This must play some role in his hostility. But most of all, I suspect, Gould thinks these ideas are dangerous and ill-motivated as well as wrong. They smack of hubris, of science moving beyond its proper domain, and incautiously at that. Dawkins does not concur. For him, knowledge of the evolutionary underpinnings of human behaviour is potentially liberating rather than dangerous. This is shown, for example, in his discussion of Axelrod's work (in the second edition of *The Selfish Gene*) on the evolution of co-operation – which he takes to be a reason for optimism about our condition.

13 Stumps Summary

Let's remind ourselves of the fundamental contrasts between Dawkins' views and those of Gould, and then summarise the state of the debate.

Dawkins argues that:

(i) Selection fundamentally acts on lineages of replicators. Most replicators are genes; chunks of DNA. But not quite all: in animals capable of social learning, some replicators are ideas or skills. And the earliest replicators were certainly not genes.

(ii) Genes typically compete by forming alliances which build vehicles. In such cases, genes succeed or fail through their distinctive, repeatable influences on these vehicles. A gene which generation by generation enhances the sensory acuity, metabolic efficiency or sexual attractiveness of a vehicle will be replicated more frequently than its rivals.

(iii) Certain genes have other replication strategies. Outlaws enhance their prospects at the expense of the adaptive design of the vehicle. Extended phenotype genes advantageously engineer the physical, biological or social environment of the vehicle they are in.

(iv) Dawkins is not committed by the logic of his position to the view that vehicles are individual organisms – groups might be vehicles. But the existence of co-operation between animals is no reason to think that groups of animals, rather than the individual animals in the groups, are vehicles. In the right circumstances, it pays individual animals to co-operate.

(v) The central problem evolutionary biology must explain is the existence of complex adaptation. So, natural selection has a special status within evolutionary biology, for complex adaptation can only be explained by natural selection.

(vi) From the perspective of evolutionary biology, humans are an unusual species. For they are vehicles of memes as well as genes. Nonetheless, the basic intellectual tools of evolutionary biology – especially those explaining co-operation, reciprocation, and sociality – apply to human evolution also.

(vii) Extrapolationism is a sound working theory. Most evolutionary patterns are accumulations over vast stretches of time of microevolutionary events. Phyla – the great lineages of animal life – began as ordinary speciation events and they grew the same way. But not quite all evolutionary patterns fit the extrapolationist view. Evolvability, for example, might involve some form of lineage-level selection.

As we have seen, Gould's picture is quite different. As he sees matters:

(i) Selection usually acts on organisms in a local population. But, in theory and in practice, selection acts at many levels. Groups of organisms can form populations of groups, with some groups doing better than others. Within a lineage of species, some may have characteristics that make them less likely to go extinct, or more likely to give rise to new species. It is even possible to have selection acting on individual genes within an organism, though this is the exception not the rule.

(ii) Many characteristics of individual organisms are not explained by selection. Furthermore, there are important patterns in the large-scale history of life that have no selective explanation. Selection is important, and evolutionary biologists must understand its operation. But it is just one of many factors explaining microevolutionary events and macroevolutionary patterns.

(iii) Extrapolationism is not a good theory. Large-scale patterns in the history of life – in particular, those tied to mass extinction episodes – cannot be understood by extrapolation from events we can measure in local populations.

(iv) Humans are evolved animals. But attempts to explain human social behaviour that make use of the techniques of evolutionary biology have largely been failures, vitiated by a one-sided understanding of evolutionary biology. They have often been biologically naïve.

These debates are still alive and developing. So, no final adjudication is possible yet. But we can say something about how the argument has developed.

The idea that gene-selectionist views of evolution are tacitly dependent on reductionism and genetic determinism is a mistake. Dawkins and the other gene selectionists do not think that nothing happens in evolution but changes in gene frequency. They do not deny the immense significance of the evolution of the organism. Rather, they see the evolution of organisms as the evolution of vehicles of selection; of 'survival machines', as Dawkins has called them. These machines interact with other survival machines, and with the inanimate environment, in a way that ensures the replication of the genes whose vehicles they are. But the construction of organisms is not the only strategy available to genes to enhance their prospects for replication. One way the gene-selectionist view of evolution contrasts with others is in its emphasis on these other strategies; on outlaws and extended phenotype genes. Extended phenotype genes are common and important, for the parasitic style of life is very common – there are millions of parasite species – and probably all parasite gene pools include genes whose adaptive effects are on host organisms. The outlaw count is unknown, but it is growing all the time. Outlaws may well turn out to be more common than we had supposed.

Gene selectionism is not genetic determinism. No gene selectionist thinks that there is typically a simple relationship between carrying a particular gene and having a particular phenotype. There *are* such genes but these are the exception not the rule. A given gene – say, the human sickle-cell haemoglobin gene – exercises phenotypic power that promotes its own replication prospects only in a given context. Change the context from one in which that gene is paired with a normal haemoglobin gene to one in which it is paired with another copy of itself, and you change the resulting phenotype.

Gene-selectionist ideas are certainly compatible with the context dependence of gene action. But they do suppose that there is some reasonably regular relationship between the presence of a particular gene in the genotype of an organism, and an aspect of that organism's phenotype. In talking about lineages of magpie aggression genes, rabbit disease resistance genes and host manipulation genes, gene selectionists assume that genes in those lineages affect their vehicles in fairly similar ways. So, while gene selectionists are not genetic determinists, they are making a bet on developmental biology. When relativised to reoccurring features of context, gene action will turn out to be fairly systematic. There is no reason to suppose this hunch is false, but it is not known to be true.

Developmental biology is salient to this debate in a second important aspect: the role of selection in evolution. Gould is betting that when the facts of developmental biology are in, it will turn out that the evolutionary possibilities of most lineages are highly constrained. The envelope of potential variation available in, say, a crustacean lineage is restricted to fairly minor alterations of that lineage's current organisation. Krill, for example, carry their gills external to their carapace, giving them a characteristic feathery appearance. Gould is betting that characters of this kind are 'frozen' into the lineage. They are developmentally entrenched. That is, these basic organisational features are connected in development to most aspects of the organism's phenotype, and that makes them hard to change. A mutation which affected the location of gills on the krill will affect many aspects of krill phenotype. Some of these effects would certainly be deleterious. For most changes in a functional system make that system less effective, not more effective. Since no variation in these frozen-in features can arise, selection is powerless to alter them and irrelevant to their persistence.

Dawkins' bets are different. Over time, selection can alter the range of evolutionary possibilities in a lineage. So he thinks both that selection has a larger range of variation with which to work, and that when patterns do persist over long periods – hundreds of millions of years – selection will have played a stabilising role. The integration of evolution and development is the hottest of hot topics in contemporary evolutionary theory, and this issue is still most certainly open.

The contrasting bets on developmental biology are still undecided, but are subjects of active research. It is harder to see how to resolve some of Gould's other claims about the large-scale history of life. Despite the great plausibility of the distinction between disparity and diversity, we are not close to constructing a good account of disparity and its measurement. Even granted that distinction, it is hard to see how to test Gould's idea that the large-scale history of life is contingent; his idea that if we 'replayed the tape' with minor variations in the starting setting, the outcome would be dramatically different. Obviously we cannot perform that experiment. And there are no natural experiments at a large enough scale. Conway Morris, in *Crucibles of Creation*, argues that evolutionary *convergence* shows that history cannot be as contingent as Gould supposes. In convergence, two independent lineages come to resemble one another when both face similar environmental pressures. Old- and New-World vultures, for example, are not closely related birds, but they are very much alike in terms of appearance and behaviour. But there are problems with this line of thought. First, most examples of convergence are not independent evolutionary experiments. For they concern lineages with an enormous amount of shared history, and hence shared developmental potential. This is true of the standard example of convergent streamlining in marine reptiles, sharks, pelagic bony fish like the tuna, and dolphins. Second, the scale is not large enough. The fact that eyes have often evolved does not show that had, say, the earliest chordates succumbed to a bit of bad luck (and become extinct), then vertebrate-like organisms would have evolved again. Third, Gould's main concern is not with adaptive complexes (the source of Conway Morris's examples) but with body plans – basic ways of assembling organisms. I think we have to score Gould's contingency claims as 'Don't know; and at this stage don't know how to find out.'

We are on somewhat surer ground with respect to Gould's ideas about high-level selection. On this issue the Dawkins–Gould divide is less sharp than it once was. For it has become clear that gene selectionism is compatible with both group selection and species selection. My bet is that Gould is right, both in thinking that mass extinctions have played a fundamental role in shaping the tree of life, and in thinking mass extinction regimes filter species in virtue of features of the species themselves, not just properties of the organisms that compose the species.

However, it has proved hard to find really clear, empirically well-founded examples to back up this hunch. At one stage, sexual reproduction was thought to be maintained by species selection. Sex has a huge cost at the individual level; a cost gene selection makes very vivid. From the perspective of every other gene in the genome, the gene coding for sexual reproduction, the sex gene, is a horrendous outlaw. For it *halves* their chances of replication in any given act of reproduction. Organisms that reproduce asexually copy all of their genes to each offspring; sexual reproducers only half. But perhaps asexual species are selected against. For as species they lack the evolutionary potential of sexual species. This idea has recently fallen on hard times. New ideas of the individual advantage conferred by sex have been developed. Moreover, the idea has a problem: sex does not always promote evolvability. For sex can break up as well as create advantageous gene combinations. If you are particularly well adapted, and you reproduce sexually, you will probably produce less well-adapted offspring. Sex will break up your particularly good gene combinations.

So it has been hard to find really convincing examples of species-level properties that are built by species-level selection. The problem is to find: (i) traits that are aspects of species, not of the organisms making up the species; (ii) traits that are relevant to extinction and survival; and (iii) traits that are transmitted to daughter species, granddaughter species and so forth. Such characteristics as a species' geographic and ecological range, its population structure and the extent of its genetic variation uncontroversially satisfy (i) and probably (ii). But are they transmitted to daughter and granddaughter species? In short, the idea that species themselves are selected is plausible, but it awaits clear confirmation.

It's time to put my cards on the table. My own views are much closer to those of Dawkins than they are to Gould. In particular, I think Dawkins is right about microevolution: about evolutionary change within local populations. But macroevolution is not just microevolution scaled up. Gould's palaeontological perspective offers real insights into mass extinction and its consequences, and, perhaps, the nature of species and speciation. So, Dawkins is right about evolution on local scales, but maybe Gould is right about the relationship between events on a local scale, and those on the vast scale of palaeontological time.

Suggested Reading

Chapter 1

Good general and impartial introductions to evolutionary theory are not thick on the ground. One very simple introduction based on Darwin himself is Ernst Mayr's *One Long Argument* (Penguin, 1991). A more technical but still readable treatment is John Maynard Smith's *The Theory of Evolution* (Cambridge, 1993). Mark Ridley's *Evolution* (Blackwell, 1996) is widely regarded as a classic. But it is demanding, and Ridley is a former student of Dawkins, so his view of the subject is influenced by Dawkins. Richard Fortey's *Life: A Natural History of the First Four Billion Years* (Vintage, 1999) is a very enjoyable narrative of evolutionary history.

Gould's review of Dennett appeared in two parts in the *New York Review of Books* in the issues of 12 June and 26 June (1997); there was a heated exchange of letters in the issue of 9 October. Gould reviewed Dawkins' *Climbing Mount Improbable* in *Evolution* (vol. 51, pp. 1020–4, 1997); Dawkins reviewed Gould's *Full House* in the same issue.

Dawkins develops his views of evolution in *The Extended Phenotype* (Oxford, 1982). This is much his best book and it is readable, though more demanding than his other works. His ideas are stated more succinctly in *The Selfish Gene* (Oxford,

1989, second edition) and developed further in *Climbing Mount Improbable* (W.W. Norton, 1996). Gould has developed his views much more through his articles than in his books. These are published both in the technical literature and in his *Natural History* column 'This View of Life'. Most of these have come out in his anthologies, but perhaps the best single presentation of his ideas is his *Wonderful Life: The Burgess Shale and the Nature of History* (W.W. Norton, 1989).

Chapter 2

The evolutionary transition from the first life-like structures to the first organisms is a hot topic in recent evolutionary theory. John Maynard Smith and Eors Szathmary's *The Major Transitions in Evolution* (Freeman, 1995) is a modern classic but tough going for the chemistry-challenged. Their *Origins of Life: From the Birth of Life to the Origins of Language* (Oxford, 1999) is a lot easier going but still somewhat demanding. Leo Buss's *The Evolution of Individuality* (Princeton, 1987) is a superb treatment of the transition from single-celled to multi-celled life. Robert Michod's *Darwinian Dynamics: Evolutionary Transitions in Fitness and Individuality* (Princeton, 1999) is an important contribution to these issues, but it is in places quite technical. Stuart Kauffman has developed a view of the origin of life which is closer to a 'cell-first' model than a 'replicator-first' model. His *The Origins of Order: Self-Organisation and Selection in Evolution* (Oxford University Press, 1993) is brutally difficult, but he returns to these themes much more gently in *At Home in the Universe* (Oxford University Press, 1995). William Schopf gives a palaeontologist's view of these issues in his: *The Cradle of Life: The Discovery of Earth's Earliest Fossils* (Princeton University Press, 1999).

Chapter 3

Dawkins' *The Blind Watchmaker* (W.W. Norton, 1986) and *Climbing Mount Improbable* (W.W. Norton, 1996) are both superb expositions of cumulative selection and its importance. The model of the evolution of the eye is from Nilson and Pelger, 'A Pessimistic Estimate of the Time Required for the Eye to Evolve', in *Proceedings of the Royal Society, B*, (vol. 256, 1994, pp. 53–8).

The idea that the genes in the egg carry the information from which the organism is built turns out to be surprisingly difficult and technical. There is no gentle introduction to this problem. I have had a go at explaining the issues in Chapter 5 of Kim Sterelny and Paul Griffiths' *Sex and Death: An Introduction to the Philosophy of Biology* (University of Chicago Press, 1999), and there is a further development of debate by John Maynard Smith's 'The Concept of Information in Biology' in the June 2000 issue of *Philosophy of Science*.

Elliott Sober presents his sceptical response to gene selection in his *The Nature of Selection* (MIT Press, 1984) and *Philosophy of Biology* (Westview, 1993). Gould develops a similar, though simpler, critique in 'Caring Groups and Selfish Genes', one of the papers in *The Panda's Thumb* (Penguin, 1980). Matt Ridley's *Genome* (HarperCollins, 2000) is an enjoyable introduction to some of the complexities of the way genes help build bodies; in this case, human bodies. (Matt Ridley is not to be confused with Mark Ridley, a student of Dawkins who also writes on evolution.)

Chapter 4

I know of no good but completely non-technical introduction to outlaw genes. But two good recent reviews are Hurst, Atran and Bengtsson, 'Genetic Conflicts', in *Quarterly Review of*

Biology (vol. 71, 1996, pp. 317–64), and Werren and Beukeboom, 'Sex Determination, Sex Ratios and Genetic Conflict', in *Annual Review of Ecology and Systematics* (vol. 29, 1998, pp. 233–61). On the role of mitochondria in turning off maleness in plants, see Saumitou-Laprade and Cuguen, 'Cytoplasmic Male Sterility in Plants', in *Trends in Ecology and Evolution* (vol. 9, 1994, pp. 431–5). These papers are written for professional audiences but they are reasonably accessible.

Dawkins himself has written the best work on extended phenotypic effects in genes, in *The Extended Phenotype*. But the three examples I give can be followed up in Gould, 'The Triumph of the Root-heads', in *Natural History* (vol. 105, 1995, pp. 10–17); Sober and Wilson, *Unto Others: The Psychology and Evolution of Altruism* (Harvard University Press, 1998) and Werren, 'Genetic Invasion of the Insect Body Snatchers', in *Natural History* (vol. 103, 1994, pp. 36–8), or, for a more recent but more technical treatment, Bourtzis and O'Neill, 'Wolbachia Infections and Arthropod Reproduction', in *Bioscience* (vol. 48, 1998, pp. 287–93).

Chapter 5

Lee Dugatkin's *Cooperation Amongst Animals: An Evolutionary Perspective* (Oxford University Press, 1997) is a good survey of the theory of animal co-operation, tied to plenty of actual examples. His *Cheating Monkeys and Citizen Bees: The Nature of Cooperation in Animals and Humans* (Free Press, 1999) covers similar material much less technically. G.C. Williams' *Adaptation and Natural Selection* (Princeton University Press, 1966) is the classic critique of group selective explanations of co-operation. Sober and Wilson's *Unto Others* (Harvard University Press, 1998) is an important attempt to revive those explanations. John Maynard Smith is responsible for applying game theory to the evolution of social behaviour. He develops this theory in *Evolution and the Theory of Games* (Cambridge University Press, 1982), but this is a technical work. Robert Axelrod's *The Evolution of Cooperation* (Basic Books, 1984) and Sigmund's *Games of Life: Explorations in Ecology, Evolution and Behaviour* are thoroughly readable alternatives. The theory of kin selection was developed by William Hamilton. These papers are in his *Narrow Roads of Gene Land*, vol. 1 (Freeman, 1996), and while the papers themselves are extremely demanding, the central ideas are explained with great clarity in his retrospectives on the papers. Gould's views on species selection are developed in two papers written jointly with Lisa Lloyd. These are: 'Species Selection on Variability' and 'Individuality and Adaptation Across Levels of Selection' both published in *Proceedings of the National Academy of Science* (vol. 90, 1993, pp. 595–9 and vol. 96, 1999, pp. 11904–9).

Chapter 6

The interaction between developmental mechanisms, variation, and evolution is one of the most active fields in contemporary research. Rudy Raff's *The Shape of Life* (University of Chicago, 1996) is a superb introduction. It is intended for a professional audience, but it is so well written (and it comes with a glossary) that it is accessible to non-experts. Wallace Arthur's *The Origin of Animal Body Plans* (Cambridge University Press, 2000) is also an important study of these issues. For an extreme version of the view that the supply of variation is so highly constrained that there is little for selection to do, see Brian Goodwin's *How The Leopard Changed Its Spots* (Charles Scribner, 1994); Goodwin's views are much more extreme than Gould's on this matter.

There is no completely non-technical introduction to the methodological issues involved in testing evolutionary hypotheses, though Griffith and I do our best to keep it simple in Chapter 10 of our *Sex and Death* (University of Chicago, 1999). The latest on these issues is canvassed in Orzack and Sober's *Adaptation and Optimality* (Cambridge University Press, 2001). The shot across the bows that began much of this is Gould and Lewontin's 'The Spandrels of San Marco and the Panglossian Paradigm: A Critique of the Adaptationist Programme', in *Proceedings of the Royal Society of London*, B, (vol. 205, 1978, pp. 581–98).

Chapter 7

As Gould sees it, extrapolationism in biology is a continuation of a cluster of methodological ideas from nineteenth-century geology known as 'uniformitarianism'. Uniformitarian ideas were formulated by Lyall, and taken up by Darwin, and hence have been part of evolutionary biology since its beginning. Gould's *Time's Arrow, Time's Cycle* (Penguin, 1988) is a fine study of uniformitarianism. Two recent and important papers on extrapolationism are: Gould, 'A Task for Paleobiology at The Threshold of Majority', in *Paleobiology* (vol. 21, 1995, pp. 1–14), and 'The Necessity and Difficulty of a Hierarchical Theory of Selection', in Anne Magurran and Robert May's *Evolution of Biological Diversity* (Oxford University Press, 1999). Jonathan Weiner's *The Beak of the Finch* is a superb exposition of selection and evolution on microevolutionary scales. That study documented season-by-season selection on finches scattered across the Galapagos islands.

Chapter 8

Eldredge's *Time Frames: The Rethinking of Darwinian*

Evolution and the Theory of Punctuated Equilibria (Simon and Schuster, 1985) is a good place to begin on punctuated equilibrium, not least because it reprints his and Gould's original article. He updates his take on the debate in *Reinventing Darwin* (John Wiley, 1995). Gould reappraises the issues as he sees them in his 'Punctuated Equilibrium Comes of Age', in *Nature* (vol. 366, 1993, pp. 223–7). Robert Carroll surveys the empirical evidence for the punctuated equilibrium pattern in his *Pattern and Process in Vertebrate Evolution* (Cambridge University Press, 1999). John Thompson, in *The Coevolutionary Process*, (University of Chicago Press, 1994) documents many examples of evolutionary changes in local populations. Elizabeth Vrba develops her 'turnover pulse hypothesis' in 'Turnover-Pulses, The Red Queen and Related Topics', in the *American Journal of Science* (vol. 293 A, 1993, pp. 418–52). Mayr's views on speciation are given most succinctly in the relevant section of his two volumes of essays, *Evolution and The Diversity of Life* (Harvard University Press, 1976) and *Towards a New Philosophy of Biology* (Harvard University Press, 1988). For a very sceptical view of punctuated equilibrium and its significance, see Daniel Dennett's *Darwin's Dangerous Idea* (Simon and Schuster, 1995).

Chapter 9

Erwin's *The Great Paleozoic Crisis* (Columbia University Press, 1993) is a superb overview of the Permian extinction. Archibald's *Dinosaur Extinction and the End of An Era* (Columbia University Press, 1996) is a judicious discussion of the debate about the most contentious extinction of them all. *Evolutionary Paleobiology*, a collection edited by Jablonski, Erwin and Lipps (University of Chicago Press, 1996), has many papers relevant to these issues. These three books are written for professional audiences, though those by Erwin and Archibald, especially, are clear and well written. Peter Ward has written two enjoyable and non-technical books on mass extinction. They are: *On Methuselah's Trail: Living Fossils and the Great Extinctions* (Freeman, 1992) and (the rather preachy) *The End of Evolution* (Bantam, 1994). Ward is sympathetic to Gould's take on these issues. Richard Fortey's *Life: A Natural History of the First Four Billion Years* (Vintage, 1999) is good on these issues, too. David Raup analyses extinction, and particularly whether mass extinction is a fair game, in his *Extinction: Bad Genes or Bad Luck?* (Oxford University Press, 1991). Gould writes on mass extinction in all of his *Natural History* collections. In these, he mostly emphasises the discontinuities in evolutionary history that mass extinction causes. He explores the idea that mass extinction imposes a filter on species, rather than the individuals that make up the species, in 'A Task for Paleobiology' in *Paleobiology* (vol. 21, 1995, pp. 1–14) and 'The Necessity and Difficulty of a Hierarchical Theory of Selection' in Magurran and May's *Evolution of Biological Diversity* (Oxford University Press, 1999).

Chapter 10

Gould makes his case about the Cambrian and its significance in *Wonderful Life: The Burgess Shale and the Nature of History* (W.W. Norton, 1989), and in it he makes a pre-emptive strike against Dawkins' line of criticism. These issues were followed up in a series of specialist articles in *Paleobiology*: Dan McShea, 'Arguments, Tests, and the Burgess Shale' (vol. 19, 1993, 399–402); Mark Ridley, 'Analysis of the Burgess Shale' (vol. 19, 1993, pp. 519–21) to which Gould replies, and especially Gould's 'The Disparity of the Burgess Shale

Arthropod Fauna: Why We Must Strive to Quantify Morphospace' (vol. 17, 1991, pp. 411–23). In these papers, Gould explores an additional way of understanding the contrast between Cambrian and post-Cambrian animal life; the distinction between a fauna with a relatively open and flexible developmental system, and a fauna with a less open, more rigid system.

Mark and Dianne McMenamin argue that selection was important in generating disparity in *The Emergence of Animals* (Columbia University Press, 1991). They suggest that the Cambrian explosion is a response to the invention of predation. Conway Morris, himself one of those who reinterpreted the Burgess Shale, takes issue with Gould in *Crucibles of Creation* (Oxford University Press, 1998). Morris argues that the history of life is much less contingent than Gould supposes, and that Gould overstates the weirdness of the Burgess fauna. In a recent review paper, Morris attempts to synthesise molecular data from the clock, and fossil evidence about both the Ediacaran and the Cambrian fauna (Morris thinks some Cambrian fauna have clear Ediacaran ancestors), to give an overview of the Cambrian explosion. See 'The Cambrian Explosion: Slow-fuse or Megatonnage', in *Proceedings of the National Academy of Science*, vol. 97, 2000, pp. 4426–9). For a good recent introduction to cladistics – the view that leads to Ridley's scepticism about the diversity/disparity distinction – see Henry Gee's *In Search of Deep Time: Beyond the Fossil Record to a New History of Life* (Free Press, 1999).

Chapter 11

The key text from Gould on these issues is *Full House* (Harmony Books, 1996), though these themes have been explored in his *Natural History* essays for years, often using baseball as a model system for discussing the importance of changes in variation in a system. Maynard Smith and Szathmary discuss evolutionary transitions in both *Major Transitions in Evolution* (Oxford University Press, 1995) and *Origins of Life* (Oxford University Press, 1999). Dawkins discusses progress both in his review of *Full House*, in *Evolution*: 'Human Chauvinism' (vol. 51, 1997, pp. 1015–20), and in 'Progress', an essay in Fox Keller and Lloyd's *Key Words in Evolutionary Biology* (Harvard University Press, 1992). Daniel Dennett defends the idea that life increases in adaptiveness over time in *Darwin's Dangerous Idea* (Simon and Schuster, 1995). J.T. Bonner defends the idea that there is a real, selection-driven trend for an increase in complexity over time in *The Evolution of Complexity By Means of Natural Selection* (Princeton University Press, 1988). Michael Ruse puts all these debates into their historical context in *From Monad to Man: The Concept of Progress in Evolutionary Biology* (Harvard University Press, 1996).

Chapter 12

Gould's views on the relationship between science and religion are explored in his *Rocks of Ages* (Ballentine, 1999). Dawkins argues for the idea that scientific knowledge is liberating in his *Unweaving The Rainbow* (Penguin, 1998). The hunch that human evolution depends as much on memes as on genes is explored most systematically not by Dawkins but by Dennett, in *Darwin's Dangerous Idea*. The best response to meme theory is Dan Sperber's *Explaining Culture* (Blackwell, 1996). For a provocative, but in my view seriously mistaken, application of evolutionary thought about humans, see Thornhill and Palmer's *A Natural History of Rape* (MIT Press, 2000). There is much better work available than this, especially that which locates human social evolution in its great-ape context. Two good recent examples are Michael Tomasello's *The Cultural Origin of Human Cognition* (Harvard University Press, 1999) and Chris Boehm's *Hierarchy in the Forest: The Evolution of Egalitarian Behavior* (Harvard University Press, 1999). Sarah Blaffer Hrdy's *Mother Nature: Natural Selection and the Female of the Species* (Pantheon Books, 1999) is also in places very speculative. But its account of the action of selection on human sexuality is much more subtle than that of Thornhill and Palmer.

Glossary

adaptation: an adaptation is a characteristic of an organism that exists today because it helped that organism's ancestors survive or reproduce.

adaptive trait: a trait that helps an organism with that trait survive or reproduce.

allele: an alternative version of a gene. Genes are located at particular regions of a chromosome. In a particular population, there may be different versions of a gene at a given location. These alternative versions are the alleles of that gene at that location.

amino acids: the building blocks of proteins. The genetic code specifies amino acids in a system that relates a sequence of three DNA bases to a single amino acid.

arms race: evolutionary interactions, within a species or between two species, in which each player becomes better adapted as a result of interaction with the other player.

biota: the totality of living things in a region or at a time.

chromosome: a long sequence of genes joined together in DNA molecules built around structurally supporting proteins. Chromosomes occur only in eukaryotic organisms. The number of chromosomes varies across species, but all (normal) members of a given species will have the same number.

clade: a lineage consisting of all of a group of species and their common ancestor.

diploid cell: a cell that has two versions of each chromosome. If the organism is the result of sexual reproduction, each parent provides one of each pair of chromosomes.

ethology: the evolutionary study of animal behaviour in the wild, rather than its study under unusual, laboratory, conditions.

eukaryotes: organisms built from complex eukaryotic cells. Each cell has a discrete nucleus, together with complex cellular machinery usually including mitochondria and, in plants, chloroplasts. Eukaryotic cells are thought to have arisen from the evolutionary fusion of bacteria-like organisms. Mitochondria and chloroplasts had free-living bacteria as ancestors.

fitness: a measure of the probability that an organism (or a gene, or a group) will reproduce itself. Comparative fitness is of particular evolutionary significance: the evolutionary history of a population will depend on which organisms (or genes, or groups) do better than others.

gamete: the sex cell of an organism (e.g. sperm, ova, pollen). It is haploid, having half the chromosome number typical of the species, and fuses in sexual reproduction with another gamete to restore the full set for the species.

gene: a DNA sequence. The exact definition of a gene remains a matter of controversy, but genes are DNA sequences of some kind. The debate is whether each gene must have an identifiable function, or whether the DNA sequences can be of arbitrary length, and with arbitrary boundaries.

genome: the total collection of genes that an organism carries.

genotype: often used as a synonym for 'genome'. But it is sometimes used to specify the genes an organism has at a specific region (or regions) of a chromosome.

haploid cell: a cell that has only a single set of chromosomes.

heritability: a measure of the probability that an offspring will share a trait possessed by its parent (at least in mathematical versions of evolutionary theory). A trait is heritable if a parent's having that trait increases the probability that its offspring will also have it.

macroevolution: a series of evolutionary changes in one or more species lineages; typically large, very long-lasting species lineages.

meiosis: the special form of cell division which generates sex cells, each of which has only half the number of chromosomes typical of cells of that species. This is in contrast to standard (asexual) cell division, whereby daughter cells end up with copies of all the structures in the parent cell.

microevolution: evolutionary changes within a single species. The term is sometimes used to refer to the evolution of one species into its immediate descendant(s).

mitochondrion: a special structure in eukaryote cells that generates energy for the cell, and has its own DNA. This DNA is almost always inherited only through the female line.

monophyletic group: a group that contains: (i) an ancestor species, (ii) only the descendants of that ancestor, and (iii) all the descendants of that ancestor.

mutation: a new DNA sequence that is produced when an error occurs in the copying process of a gene (or another replicator), resulting in a difference between the daughter gene and the template from which it was copied. Mutations are one source of new genetic variation in the population. Most, if they have any effects, have bad ones. So selection has acted to make the copying process very accurate indeed. But organisms have so many genes that even accurate copying still generates appreciable numbers of mutations.

natural selection: the process by which the superior fitness of certain traits causes those traits to increase in frequency in a population.

phenotype: an organism's developed morphology, physiology and behaviour. It contrasts with the genotype: the genes that an organism carries.

prokaryotes: single-celled organisms, such as bacteria, without a nucleus or mitochondria. Prokaryotes are the simplest and oldest forms of life.

protein: a very large molecule made up of chains of amino acids folded in extraordinarily complex ways.

replicator: a structure that causes copies of itself to be made, and that, in combination with others, sometimes constructs a vehicle of selection. It is Dawkins' unit of heredity and selection.

species: there is no uncontroversial definition of a species. The most usual definition is the 'biological species concept' which defines a species as an interbreeding population of organisms. But there are many problems in making this notion precise. Moreover, using this definition, no asexual organisms form species.

species sorting: any pattern in species survival or extinction counts as species sorting, whatever the cause of that pattern. If, for example, for whatever reason, species with small population sizes are at a greater risk in mass extinction events, that would count as species sorting.

vehicle: a structure built by gene combinations in development. A vehicle mediates the reproduction of the genes responsible for its production. The clearest examples of vehicles are individual organisms, but there may be others, including groups of organisms.

Appendix: Geological Time Scale

Era	Period	Epoch	Duration
Cenozoic	Quaternary	Holocene Pleistocene	100,000 BP to present 2mya–100,000 BP
	Tertiary	Pliocene Miocene Oligocene Eocene Palaeocene	5–2 mya 24–5 mya 38–24 mya 55–38 mya 65–55 mya
Mesozoic	Cretaceous Jurassic Triassic		144–65 mya 213–144 mya 248–213 mya
Palaeozoic	Permian Carboniferous Devonian Silurian Ordovician Cambrian		286–248 mya 360–286 mya 408–360 mya 438–408 mya 505–438 mya 590–505 mya
Precambrian	*Various*		4,600–590 mya

BP = years before present
mya = millions of years ago
All figures are approximate.